Lecture Notes in Electrical Engineering

Volume 200

For further volumes:
http://www.springer.com/series/7818

Society of Automotive Engineers of China
(SAE-China) · International Federation of
Automotive Engineering Societies (FISITA)
Editors

Proceedings of the FISITA 2012 World Automotive Congress

Volume 12: Intelligent Transport System (ITS) & Internet of Vehicles

Editors
SAE-China
Beijing
People's Republic of China

FISITA
London
UK

ISSN 1876-1100 ISSN 1876-1119 (electronic)
ISBN 978-3-642-33837-3 ISBN 978-3-642-33838-0 (eBook)
DOI 10.1007/978-3-642-33838-0
Springer Heidelberg New York Dordrecht London

Library of Congress Control Number: 2012948289

© Springer-Verlag Berlin Heidelberg 2013
This work is subject to copyright. All rights are reserved by the Publisher, whether the whole or part of the material is concerned, specifically the rights of translation, reprinting, reuse of illustrations, recitation, broadcasting, reproduction on microfilms or in any other physical way, and transmission or information storage and retrieval, electronic adaptation, computer software, or by similar or dissimilar methodology now known or hereafter developed. Exempted from this legal reservation are brief excerpts in connection with reviews or scholarly analysis or material supplied specifically for the purpose of being entered and executed on a computer system, for exclusive use by the purchaser of the work. Duplication of this publication or parts thereof is permitted only under the provisions of the Copyright Law of the Publisher's location, in its current version, and permission for use must always be obtained from Springer. Permissions for use may be obtained through RightsLink at the Copyright Clearance Center. Violations are liable to prosecution under the respective Copyright Law.
The use of general descriptive names, registered names, trademarks, service marks, etc. in this publication does not imply, even in the absence of a specific statement, that such names are exempt from the relevant protective laws and regulations and therefore free for general use.
While the advice and information in this book are believed to be true and accurate at the date of publication, neither the authors nor the editors nor the publisher can accept any legal responsibility for any errors or omissions that may be made. The publisher makes no warranty, express or implied, with respect to the material contained herein.

Printed on acid-free paper

Springer is part of Springer Science+Business Media (www.springer.com)

Contents

Part I　Driver Assistance System

Development of Lane Detection System Based on FPGA........... 3
F2012-I01-002
Feng Gao, Hanyang Gui, Qiang Zhang and Jugang He

**Approaches of User-Centered Interaction Development
for Highly Automated Vehicles in Traffic-Jam Scenarios**.......... 13
F2012-I01-006
Felix Wulf, Maria Rimini-Doering, Marc Arnon and Frank Gauterin

Integrated Simulation Toolset for ADA System Development....... 25
F2012-I01-007
Martijn Tideman, Roy Bours, Hao Li, Tino Schulze and Takahito Nakano

**Haptic Velocity Guidance System by Accelerator Pedal Force Control
for Enhancing Eco-Driving Performance**...................... 37
F2012-I01-012
Feilong Yin, Ryuzo Hayashi, Raksincharoensak Pongsathorn
and Nagai Masao

**Physical Model-Based Yaw Rate and Steering Wheel
Angle Offset Compensation**............................... 51
F2012-I01-014
Stefan Solyom and Johan Hultén

Dominant Driving Operations in Curve Sections Differentiating Skilled and Unskilled Drivers 59
F2012–I01-016
Shuguang Li, Shigeyuki Yamabe, Yoichi Sato, Takayuki Hirasawa, Suda Yoshihiro, P. N. Chandrasiri, Kazunari Nawa, Takeshi Matsumura and Koji Taguchi

Cross-Linking Driver Assistance Systems Via Centralized Scene Interpretation Using The Example Of Directional Roadway Detection 73
F2012-I01-018
Markus Hörwick, Loren Schwarz, Stefan Holder, Martin Buchner and Hariolf Gentner

Part II   V2X Communication Technology

Towards Design and Integration of a Vehicle-to-X Based Adaptive Cruise Control 87
F2012-I02-002
Oliver Sander, Christoph Roth, Benjamin Glas and Jürgen Becker

Traffic Signal Information in a Real Residential Area 101
F2012-I02-004
Benno Schweiger, Regina Glas, Christian Raubitschek and Johann Schlichter

V2X Communication Technology: Field Experience and Comparative Analysis 113
F2012-I02-005
Heri Rakouth, Paul Alexander, Andrew Jr. Brown, Walter Kosiak, Masao Fukushima, Lali Ghosh, Chris Hedges, Henry Kong, Sven Kopetzki, Ramesh Siripurapu and Junqiang Shen

Improved System for ParkNet Mobile Network 131
F2012-I02-006
Yao Ge, Wenzhi Xue and Zhan Shu

Application Perspectives for Active Safety System Based on Internet of Vehicles 147
F2012-I02-007
Ling Chen

Contents

Part III Telematics, Navigation System

Consideration on the Time-Based Evaluation of the Traffic Information Prediction 155
F2012-I03-002
Ishiguro Yosuke

Barge-in Implementation Method for Multi-CPU In-Vehicle Speech Recognition System 163
F2012-I03-004
Naoyori Tanzawa and Yoshikazu Inagaki

Seamless Traffic Information in Dynamic Navigation Up to Date 169
F2012-I03-005
Regina Glas, Heidrun Belzner, Tim Lange, Irina Koller-Matschke and Richard Wisbrun

Traffic Information and Individual Driver Behaviour 179
F2012-I03-007
Irina Koller-Matschke, Heidrun Belzner and Regina Glas

Smart Automotive Apps: An Approach to Context-Driven Applications 187
F2012-I03-010
Stephan Durach, Uwe Higgen and Michael Huebler

Pre- and Postdrive Predictions 197
F2012-I03-013
Carsten Isert and Oliver Stamm

Part IV Eco Driving Technology

Adaptive Cruise Control: A Behavioral Assessment of Following Traffic Participants Due to Energy Efficient Driving Strategies 209
F2012-I04-002
Dirk Hülsebusch, Maike Salfeld, Yinchao Xia and Frank Gauterin

System-Oriented Validation Aspects of a Driver Assistance System Based on an Accelerator-Force-Feedback-Pedal 221
F2012-I04-003
Albert Albers, Alexander Schwarz, Christian Zingel, Jens Schroeter, Matthias Behrendt, Andreas Zell, Carmelo Leone and Antonio Arcati

Part V Harmonization and Regulation of ITS Systems

Dynamic Optimal Model Researches for Correlative Intersections Control Based on Particle Swarm Optimization Aiming at Multi-Subsystem 237
F2012-I05-002
Yifeng Huang and Song Luo

Measurement of Electrodermal Activity to Evaluate the Impact of Environmental Complexity on Driver Workload............... 245
F2012-I05-003
Maria Seitz, Thomas J. Daun, Andreas Zimmermann and Markus Lienkamp

Part VI Other

Study on Engine Idle Speed Control Considering Vehicle Power Balance 259
F2012-I06-002
Feng Gao, Qiang Zhang, Daquan Zhang and Wen He

Intelligent Functionalities for Fully Electric Vehicles 267
F2012-I06-005
Adrian Zlocki, Qihui Huang, Mimoun Ghaouty El, Lutz Eckstein and Holmer-Geert Grundmann

Motion Stabilizing Controller of Off-Road Unmanned Wheel Vehicle in 3 Dimensional Space 275
F2012-106-006
Yue Ma, Changle Xiang, Qingdong Yan and Quanmin Zhu

Part I
Driver Assistance System

Development of Lane Detection System Based on FPGA

Feng Gao, Hanyang Gui, Qiang Zhang and Jugang He

Abstract Lane detection is the key part of LDW. In this paper a lane detection algorithm and the corresponding FPGA circuit were developed based on Spatran6 platform of Xlinx. Several ingenious structures are used to improve processing speed when designing FPGA circuit. Its performances were validated by co-simulation of software and hardware. Results show that the detection accuracy is 100 % at normal condition when entering tracking mode. The processing speed is more than 30 frames/s and it can also be used in especial conditions, such as dark, curve, etc.

Keywords Machine vision · Lane departure warning · Field programmable gate array · Lane detection · Co-simulation

1 Introduction

In order to improve the safety of vehicle, many active safety systems based on machine vision technology, such as Lane Departure Warning System (LDW), etc. have already been in mass production [1–4]. Digital Signal Processing Unit (DSP) has been widely used in these systems. It is the trend that more and more active safety systems based on vision technology will be applied in vehicle and they have to be integrated together to reduce cost and risk. At that time, only DSP can hardly

---

F2012-I01-002

F. Gao (✉) · H. Gui · Q. Zhang · J. He
Automobile Engineering Institute of Changan Automobile Ltd. Co, Chongqing, People's Republic of China

satisfy the real time requirement of active safety systems. It is a trend that Field Programmable Gate Array (FPGA) will be applied in image and video processing system, which can increase the processing speed greatly because of its parallel process.

Lane detection is one of the most important parts of LDW, which always consumes lots of DSP resources [5–7]. In order to improve the processing speed of LDW and integrate more active safety functions into one system, a lane detection algorithm based on FPGA is designed in this paper. Several first input first output (FIFO) and parallel channel structures are used. By this way, much more data can be processed at the same time. The designed lane detection FPGA circuit is generated by the System Generator of ISE, which is an integrated development tool for FPGA and produced by Xilinx. The performances of the designed lane detection method are validated by co-simulation on the Spatran 6 development platform. The test results show that the performances of real time, lane detection accuracy and the complex traffic condition adaptability of the designed FPGA circuit for lane detection can meet the requirements of LDW system.

2 Lane Detection Algorithm

The main structure of the detection algorithm is depicted in Fig. 1.

The lane detection algorithm includes two working modes. One is initial model and the other is tracking mode [6, 7]. The mode is selected by detector state controller according to the processing results. In order to reduce the processing data, in initial mode, only data in area II and III is processed and in tracking mode only data in the interesting area is processed. The definition of the image area is shown in Fig. 2.

In the edge detection/binarization block, Sobel arithmetic operator is used to detect object edge and then the threshold value used in binarization is derived by Otsu's Thresholding Method [8]. For every pixel, whose coordinates are (x_i, y_i), parameter θ is traversed and the corresponding parameter ρ can be derived by Eq. (1). The calculated results are added up at a two dimension space, whose coordinate axes are ρ and θ respectively. The added up value represents the count of dots in the line, whose angle and distance are corresponding value of θ and ρ respectively. To reduce the calculation load, considering the possible position of the lane, the angle θ of left line is traversed between 15 and 90° and that of right lane is traversed between −15 and 90°.

$$x\cos(\theta) + y\sin(\theta) = \rho \quad (1)$$

Based on the available lane detection algorithms, to improve the robustness of lane detection, the "Lane Pick-up" and "Detector State Controller" blocks are redesigned in this paper.

Development of Lane Detection System Based on FPGA

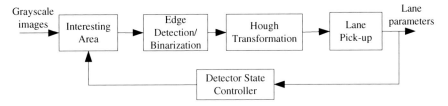

Fig. 1 Lane detection algorithm

Fig. 2 Image area definition

2.1 Lane Pick-Up Module

A matrix is produced by "Hough Transformation" block. Each element in the matrix represents a line and the value equals the count of dots in this line. The objective of "Lane Pick-up" block is to find the effective lane parameters from the above matrix. During initial mode, the pixels, which comprise other objects aside road such as electricity pole, may be much more. Picking the element whose value is biggest as effective lane simply may result in wrong result. And so, a new algorithm is developed as shown in Fig. 3 to pick up effective line from Hough transformation matrix.

The threshold is derived by multiplying the matrix biggest value with a factor, which makes the system have the adaptability to different conditions. The values around the matrix biggest value are also much bigger than others, so after finding the biggest value the around values are reset in order to avoiding their influence on finding other peak values. In this paper, three peak values are found out to. Then the angle of the three lines is used to judge its effectiveness. The line whose slope is bigger is picked up as effective line.

Fig. 3 Lane pick up method

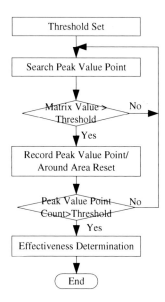

2.2 Detector State Controller

In real driving condition, the position of the line varies very little between two continuous images. It means that only the area near the line detected in the last cycle should be processed. So the lane detection algorithm has two modes. One is initial mode and the other is tracking mode. In different mode, the interesting area is different. The function of "Detector State Controller" module is to determine the working mode.

2.2.1 Switching from Initial Mode to Tracking Mode

In initial mode, objects on the road greatly disturb the lane detection results, such as texts, direction arrows, etc. If the input lane parameters of "Detector State Controller" are wrong, the correct lane will not be found in the following process. So before switching to tracking mode, the lane detection should be stable enough. If the correct lane is found, the lane parameters will not vary much between two continuous images. So the stability of lane detection is determined by the following procedure.

Step 1: Calculate the differences between lane parameters of the two continuous images.
Step 2: Stored the calculated above value into a FIFO. Continuous ten values are stored in this paper.
Step 3: If the amount of the values, which are bigger than a predefined value, is smaller than the threshold, lane detection is stable.

Only lane detection is stable is not enough, since there are lots of disturbances, such as direction arrows, zebra crossing, etc., which will appear in several continuous images. Besides the above criteria, another one is added, which is whether the distance between the detected two lines satisfies Chinese standard. To simplify calculation, the distance is calculated along a predefined horizontal position in the image. If the distance lies in the reasonable range, the detected two lines will be considered as lanes.

Both above two conditions are satisfied, the lane detection algorithm switches from initial mode to tracking mode.

2.2.2 Switching from Tracking Mode to Initial Mode

In real driving condition, dirty lane, discontinuity of lane, etc. will result in temporary error of lane detection. If detection algorithm switches from tracking mode to initial mode immediately when detection error occurs, the system will switch between two modes frequently. So a switching delay algorithm is designed in this paper.

When the difference between lane parameters is bigger than the threshold or no effective lane is found, one is added to the counter. The interesting area of the last image is used in the next image process and the algorithm stays in tracking mode. Otherwise one is subtracted from the counter. The algorithm switches from tracking mode to initial mode when the value of the counter is bigger than the threshold.

3 FPGA Circuit Design

The structure of the designed FGPA circuit for lane detection is shown in Fig. 4.

The designed FPGA circuit mainly consists of two parts. One is the image precondition and the other is the lane detection. The objective of image precondition circuit is to wipe off the disturbances of background and derive the useful binarization image data. The function of the lane detection circuit is to find out the effective lane parameters from the above binarization image.

3.1 Image Precondition Circuit

A FIFO structure is used as a data cache to store the image data, so three line data can be read at the same and the gradient of each pixel can be calculated in parallel. The calculated gradient data is stored in another FIFO structure, which makes the following edge thinness process circuit work in parallel.

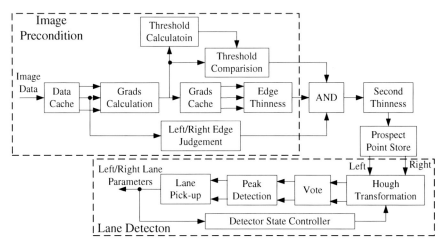

Fig. 4 FPGA circuit for lane detection

After gradient calculation, all the value will be added together. When finishing processing all data of an image, the sum of the gradient is obtained at the same time. It will be used to calculate the gradient threshold of next image.

To reduce the calculation of "Hough Transformation", an "Edge Thinness" circuit is added before "Lane Detection", which makes the width of the lane image to one pixel. The "Prospect Point Store" is to store the prospective points of left down image and right down image separately.

3.2 Lane Detection Circuit

Two same process channels are used for lane detection, so the right and left image can be processed at the same time. X and Y axis values of Hough transformation matrix are calculated offline and stored in registers. By this way, the calculation of sine and cosine is avoided.

The function of "Vote" is to read out the value in the corresponding memory unit according to the line parameters, add one to it and then write it back to the memory unit. In the initialization mode, the "Vote" circuit is activated by "Hough Transformation". In the tracking model, the enable condition of the "Vote" is controlled by both "Hough Transformation" and "Detector State Controller". When "Detector State Controller" finds that some pixel is not in effective tracking area, the "Vote" will be disabled.

During Hough transformation, much data should be stored and the peak value search can only be performed after transformation normally. This leads to the need of more registers and process time. A special circuit, which is similar to a shift register, is designed to search the peak value in Hough transformation matrix.

Development of Lane Detection System Based on FPGA

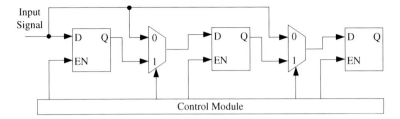

Fig. 5 Peak value detection circuit

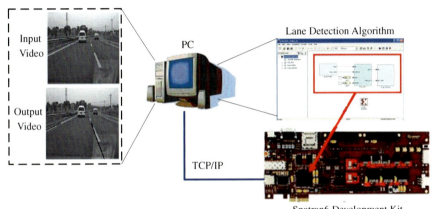

Fig. 6 Software–hardware co-simulation system

3.2.1 Peak Value Detection

The special peak value detection circuit is shown in Fig. 5.

The circuit has three flip-latches and the data stored in the right latch is smaller than that in the right. The characteristics of this circuit are:

1. When all enable signals are set, the output of left latch is the input of right. The circuit becomes a shift register.
2. When some data need to be stored into a certain latch, its enable signal is set. By this, the data can be stored and shifted in one cycle.

The advantage of this design is that the peak value detection is done like a pipeline. When Hough transformation is finished, the peak values are found at the same time.

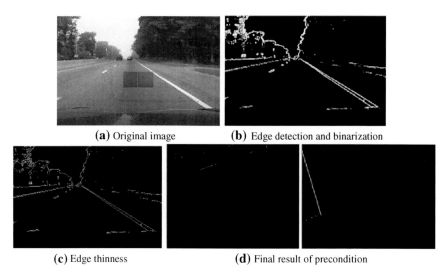

Fig. 7 Image precondition results. **a** Original image. **b** Edge detection and binarization. **c** Edge thinness. **d** Final result of precondition

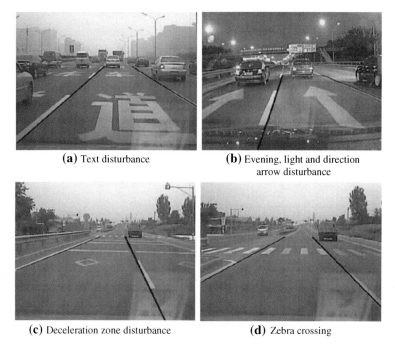

Fig. 8 Lane detection results in different traffic condition. **a** Text disturbance. **b** Evening, light and direction arrow disturbance. **c** Deceleration zone disturbance. **d** Zebra crossing

4 Software–Hardware Co-Simulation

To validate the effectiveness of the designed circuit, Spatran 6 development kit of Xlinx is used to do software-hardware co-simulation. The co-simulation system is shown in Fig. 6.

The test video is from real traffic, which includes different kinds of disturbances, such as evening, zebra crossing, text, crossover, deceleration zone, vehicle, light, etc. The raw data of test video and the processed video is transferred between computer and Spatran 6 development kit through TCP/IP protocol. Figure 7 is some results of the image precondition. From the results, it can be found that the proposed process method can clear up disturbances from background. Figure 8 is the lane detection results in different road conditions. In the tracking mode, the process speed is 30 frame/s and the detection accuracy is 100 % if the lane is visible. The designed lane detection circuit is robust to disturbances such as text, zebra crossing, etc.

5 Conclusions

A lane detection algorithm based on FPGA was designed in this paper. The FPGA circuit has been validated by software-hardware co-simulation. The following conclusions can be derived from the results.

1. The designed image precondition method can remove disturbances of background. The lane information is reserved, while the amount of data to be processed is reduced as much as possible.
2. The processing speed of the developed lane detection FPGA circuit is 30 frame/s based on Spatran 6 development kit of Xlinx.
3. The lane detection system is robust to disturbances of text, zebra crossing, vehicle, etc. In tracking mode, the lane detection accuracy is 100 % under the test conditions.

References

1. Liu J, Bin H (2007) Lane departure warning system based on monocular vision. Electron Eng Prod World 3:128–130
2. Yuan Q (2010) Traffic lane deviation early warning system design based on TMS320DM642. Manuf Autom 5:70–72
3. Wu M (2005) Research on approach for vision based lane departure warning system. National University of Defense Technology, Changsha
4. Pingshu GE (2008) Study on the modified vision algorithm of lane departure warning system. Jilin University, Changchun
5. Li X, Zhang W (2008) Research on lane departure warning system based on machine vision. Chin J Sci Instrum 7:1554–1558

6. Yu T (2006) Study on vision based lane departure warning systems. Jilin University, Changchun (in Chinese)
7. Dong Y (2004) Study on algorithm of automotive highway lane departure warning system. Jinlin University, Changchun
8. Rafael CG, Richard EW, Steven LE (2005) Digital image processing. Publishing House of Electronic Industry, Beijing

Approaches of User-Centered Interaction Development for Highly Automated Vehicles in Traffic-Jam Scenarios

Felix Wulf, Maria Rimini-Doering, Marc Arnon and Frank Gauterin

Abstract In the recent years one of the goals of driver assistance systems has been to disburden the driver of parts of the driving task. Current developments are able to take over full control of the vehicle in specific use cases. One of the main challenges is to maintain the driver's ability to take back the responsibility while being driven by the car. As it is assumed that the driver's main benefit of such systems is the possibility to perform secondary tasks, a target conflict comes up. This paper describes a systematic approach to derive new ways of interaction between the driver and the automated vehicle. Traffic jam situations are chosen as the example situational context for this paper since they are one of the first expected use cases for automated driver assistance systems. By incorporating knowledge of experts of interdisciplinary fields in an innovation workshop, multiple points of view towards the given problem are enabled. For good rated ideas the generalized functional principle is formulated which describes how the user is kept aware of his supervision task. The resulting functional examples are decoupled from the initial ideas by deriving them only from these principles.

Keywords Driver assistance · Automation · Traffic jam · Human–machine interaction · Driver attentiveness

---

F2012-I01-006

F. Wulf (✉) · M. Rimini-Doering · M. Arnon
Robert Bosch GmbH, Stuttgart, Germany

F. Gauterin
Karlsruhe Institute of Technology (KIT), Karlsruhe, Germany

1 Introduction

Due to the increasing mobility, there is a growing demand for comfort and safety in traffic. Current developments in the sector of driver assistance systems (DAS) concentrate on the lateral guidance of vehicles (e.g. [4] or [13]). The resulting automatic steering control (ASC) is able to take over lateral control of the vehicle. The next step will be the combination of ASC with existing systems like adaptive cruise control (ACC), which is capable of taking over longitudinal control of the vehicle [14]. The emerging integrated cruise assistance systems (ICA) will therefore be able to take over longitudinal and lateral control of the vehicle in specific environments and hence disburden the driver of parts of his driving task. Stanton and Marsden [12] have identified significant potential benefits of automated driving functions:

- Improved well-being of the driver,
- Reduction of driving error and therefore enhancement of safety,
- Greater attractiveness and desirability of such driver assistance systems and therefore increased sales.

One of the first expected use cases for automated DAS is driving in traffic jams. This is due to the fact that the driver is usually underloaded in traffic jams because of the tiresome and monotonous driving situation. Hence, this is one of the scenarios where good advantages can accrue from automation.

The following chapters describe the human role in an automated vehicle. This motivates a need for new interactions schemes for automated DAS. Afterwards an overview on an innovation process is given. It is designed to generate new interaction schemes for automated driver assistance systems. Subsequently, the process is conducted using the example of automated driving in traffic jams.

2 The Human in Automated Vehicles

Recent research [5] has identified several degrees of automation to characterize driver assistance systems that can take over both longitudinal and lateral control of the vehicle. The main distinctive features for the characterization of such systems are the necessity for permanent observation of the automation and the necessity for the driver's constant ability of controlling the vehicle. If both needs are given, the corresponding level of automation is called "partly automated". In the higher levels of automation permanent attention of the driver is not necessary which leads to legal issues covering product liability and the question of responsibility in case of an accident. Since these issues are not yet solved, partly automatic DAS are seen as the next step towards autonomous driving. The current paper deals only with partly automatic DAS.

As mentioned above, in partly automatic DAS the driver needs permanent attention. The attention level can be represented by the concept of situation awareness. Endsley [3] defines situation awareness as the "perception of the elements in the environment within a volume of time and space, the comprehension of their meaning and the projection of their status in the near future". Consequently, one can speak of three levels of situation awareness (perception, comprehension and projection). A sufficient global level of situation awareness is needed in order to react adequately to external stimuli. Rauch [11] states that automation can have a negative effect on situation awareness. In case of system errors, a lack of situation awareness can lead to late, wrong or even missing counter-reactions of the operator [2, 8].

A sufficient level of situation awareness is essential. Hence, the fulfilment of this requirement is one of the main aims in the current research activity. This paper focuses on the process of discovering new interaction schemes between the driver and a partly automated DAS which can support situation awareness.

The term "interaction" denominates in this case the collaboration of the driver and the automated system in order to perform the task of driving the vehicle. This can mean both the allocation of drive-related tasks between human and automation and the communication between them e.g. through displays or operational controls.

Any new interaction element has an influence on the characteristics of the DAS. This can potentially also change the acceptance and the perceived benefit of the end user. Both are among the main criteria for the success of DAS. Therefore, these aspects are also taken into account in the evaluation of the interaction schemes.

3 Overview on the Process of Deriving Interaction Schemes to Support Situation Awareness

In order to find new interaction elements, a specific innovation process was conducted. The development was based on the process to find new products, described by Pahl and Beitz [9]. Two steps were added in order to enhance the quality of the results. Thus, it consists of four main stages which are visualized in Fig. 1.

In the first step, new ideas are generated. It is therefore called the Generation step. According to the basic process [9], new thoughts can be found through various methods. One of them is the innovation workshop using intuitive methods such as a brainstorming or other creativity techniques. The aim of this step is to generate the highest possible number of ideas.

After the generation of new ideas, they must be evaluated and filtered in the second step of the process (Filtering). Therefore, similar ideas are identified and merged. Afterwards the effectiveness of every idea towards the given criteria of

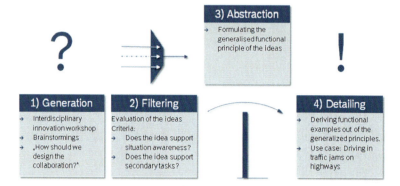

Fig. 1 Overview on the approach of generating generalized functional principles and examples

the innovation process is rated. The objective of this step is to reduce the quantity and to raise the quality of the resulting items.

It is supposed that many of the resulting ideas are designed according to few functional principles. This stands in analogy to the TRIZ-method, which states that many technical conflicts can be solved using only a few abstract principles of invention [6]. Thus, in order to get a better overview on the resulting ideas, the Abstraction step is added to the standard process described by Pahl and Beitz [9]. Every good rated element is generalized and its underlying functional principle is formulated. This principle describes how the given goals shall be reached. This is a further step in order to reduce the quantity and to raise the quality of the resulting items.

In the subsequent Detailing step, the resulting elements of the process are derived only from these generalized principles. The advantage of this procedure is that the results are decoupled from the conceived ideas in the first step and that new perspectives on the same principles can be allowed. Thus, the results become more holistic.

In the following chapters, the process is conducted using the example of automated driving in traffic jams.

4 The Generation Step

In the generation phase, a workshop with 14 experts was conducted. In order to ensure multiple points of view towards the given problem, the participants were recruited from three different groups:

- Experts in the field of driver assistance systems,
- Experts in the field of Human–Machine Interaction,
- Various other technical experts.

Approaches of User-Centered Interaction Development

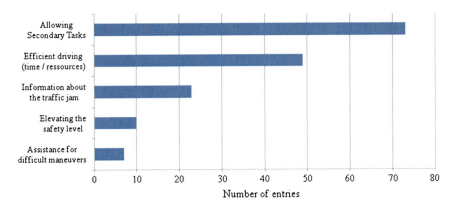

Fig. 2 End-user benefits of DAS in traffic jams

The workshop was conducted by a team of two facilitators. It was divided into two parts, in each of which a brainstorming with a different focus was carried out.

The first brainstorming concentrated on the end user benefit of driver assistance systems in traffic jams in general. Therefore the participants were asked how they would wish to be assisted while driving in congested traffic. On the one hand, the resulting ideas contained general categories of end user benefit. On the other hand, specific interaction schemes were mentioned, which were sorted into the general benefit categories according to their specific scopes. Figure 2 shows the distribution of entries to the resulting benefit categories.

The results show that the end user benefit expected by most participants is the ability to perform secondary tasks while being driven by the system. The tasks mentioned by the participants cover a wide range of activities e.g. reading news or playing games. Hence, the ability to perform secondary tasks shall be the second requirement which adds to the first requirement, keeping a sufficient level of situation awareness. Secondary tasks can have negative effects on situation awareness [10]. Thus the two requirements can possibly be in conflict with each other. Hence, there is need for interaction schemes that on the one hand enable the driver to perform secondary tasks but on the other hand ensure that he has a sufficient level of situation awareness. This underlines the need for the present research activities.

The second important end user benefit of such a DAS in traffic jams is efficient driving. By reduction of driving errors concerning the choice of speed and distance, the automation is able to smooth the traffic flow in jams raising the average speed and thus reducing time losses and fuel consumption. This is a requirement towards the application of the DAS. It is therefore not specifically related to the driver's attentional state. The remaining mentioned benefits are either part of other systems like navigation devices ("Information about the traffic jam") or do neither have specific relations to the driver's attention ("Elevating the safety level through reduction of driver errors" and "Assistance for difficult maneuvers"). Therefore these requirements are not in this paper's focus.

Before the start of the second brainstorming the participants were informed about technical details of future partly automatic driver assistance systems and about the constant need for the driver to supervise the system and to intervene in case of danger. The brainstorming was focused on the question how driver and system have to cooperate in order to keep the driver in charge of the driving process while he or she is at the same time being driven completely by the system. This was the basic idea generation part in which new interaction schemes should be generated. The participants were divided into five groups. Each of these groups concentrated on certain situations that can occur using an automated DAS in traffic jams, e.g. "turning the system on and off" or "behavior in case of special infrastructure".

5 The Filter Step

The second step of the innovation process is the filter. In general its goal is to reduce the number of different ideas that have to be considered. Hence, similar ideas have to be merged and the resulting items have to be evaluated. Those ideas which do not contribute to the given goals have to be identified and eliminated. As stated above, there are two main requirements which potentially oppose each other. On the one hand, the driver has to maintain a sufficient level of situation awareness. But on the other hand, the driver's benefit of the automation shall be raised by enabling him to perform secondary tasks during the drive in traffic jams.

As some of the ideas emerging from the first brainstorming also contain specific interaction schemes, both conducted brainstormings are seen as input for this step. First, similar ideas coming from both brainstormings are merged. The subsequent evaluation is performed by a team of experts in the two fields of "Human–Machine Interaction design" and "development of DAS". Each item emerging from the innovation workshop is reviewed considering the two given goals. There are three categories of evaluation for each element and goal:

- The item has a negative effect on the given goal,
- The item has a neutral effect on or is irrelevant for the given goal,
- The item has a positive effect on the given goal.

Items that do not have any positive effect on any of the goals are neglected. By performing the filter step, the amount of items could be further narrowed.

6 The Abstraction Step

For each item resulting from the previous filter step, a generalized functional principle is formulated. As related ideas can have the same underlying principle, this is a further step reducing the amount of different items to be considered.

Approaches of User-Centered Interaction Development

The functional principle should comprise how the given goal shall be reached. The following explanations give a brief introduction to the found principles. They can be on the one hand grouped into principles concerning the allocation of the driving task between the driver and the automation and on the other hand into principles concerning the communication schemes between the driver and the automation. In the following, the resulting four principles are described shortly.

1. Artificial degradation of the systems performance (limiting the quality of control)

This principle is based on the fact that the underlying DAS either does not take over the driving task completely or that the quality of its vehicle control is artificially reduced compared to a fully functional system. For example some special driving situations may be intentionally executed badly. This system behaviour must be clearly shown to the driver through frequent appearance. By doing that, the driver will become well aware of the system's limits and does not rely too much on it. Hence, the driver will stay involved in the task of driving and a sufficient level of situation awareness is kept. Too high frequencies can have negative effects on the driver's trust to the system [2] and therefore affect the user acceptance. This principle follows.

2. Artificial degradation of the system's availability (limiting the quantity of control)

By reducing the availability of the DAS, the ratio of automated driving is reduced. As the driver has to take over control in a relatively high number of situations, he is always prepared for doing so. Hence the level of situation awareness stays high.

3. Guidance of driver's attention towards the surrounding environment

According to Endsley's model of situation awareness [3], the logical first step in order to enhance the global level of the driver's situation awareness is to support him in the perception of his environment. To do so, his attention must be guided geometrically into the directions that are important at the moment. As the perception during the driving task is performed up to 90 % using the optical channel [1], special importance must be given to the direction of view.

4. Monitoring the driver's attention towards the surrounding environment

The attentional state of the driver can be seen as an additional system boundary. As soon as the driver looses the needed level of situation awareness, he can be warned and the automation can be shut down with an adequate intervention strategy. One of the main issues is the detection of the driver's state. Rauch et al. presented an approach to identify the driver's state using a camera based system [10]. As there is still an intensive demand for development of such driver state assessment systems, this generalized mechanism is not in focus of this work.

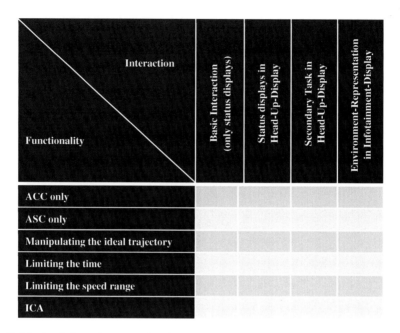

Fig. 3 Matrix of alternatives emerging from the innovation process

7 The Detailing Step

The last step in the given process is the derivation of specific realizations. Consequently, another brainstorming was conducted with four experts in the field of driver assistance. The given task was to find solutions, which are designed according to the given generalized principles from the preceding step. Thus, the results are independent from the output of steps one and two and the amount of point of views to the problem can be further raised.

Due to the additional step conducted around the generalized functional principles, the resulting list contains several new items (20 %). This states that the developed process helped to gain additional views onto the given problem. Thus, the results become more holistic.

Every driver assistance system consists of a functionality scheme and a communication scheme. Consequently, the resulting functional examples must comprise at least one element from each group. The resulting functional examples can thus be arranged in form of a matrix (Fig. 3). As mentioned above, principle 4 is not in focus of the current work and therefore not represented in the matrix.

Functionalities which are designed according to principle 1 reduce the system's performance in relation to an ICA. Trivial elements are therefore the existing systems ACC or ASC in a stand-alone package. Either the lateral or the longitudinal control is not taken over by the system which can be seen as a reduction of performance in relation to an integrated system. Another way to reduce the

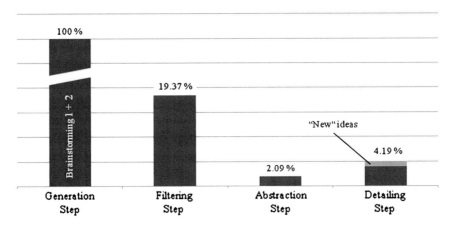

Fig. 4 Amount of items resulting from the various process steps

performance of the system is to manipulate the lateral and/or longitudinal trajectory in relation to the ideal trajectory. By doing that, mistakes are built in. It must be ensured that these mistakes do not lead to dangerous situations. One example of this is a system that constantly drives a little bit slower than necessary to follow the preceding vehicle. By doing that, the gap between the vehicles will be growing steadily and the driver has to intervene occasionally.

Functionalities designed according to principle 2 reduce the systems availability. This can be done by limiting either the time chunk the automation may be turned on without driver interaction, the speed range or by a combination of both.

The resulting communication elements are all based on principle 3, the guidance of the driver's attention towards the surrounding environment. This is done by arranging visual elements the driver needs or wants to watch (e.g. system status indicators or secondary tasks) together with the surrounding environment. A trivial solution for this is placing them in a head-up-display. A possible realization of this is described in [7]. Another way is to place visual elements that represent the environment (e.g. video) in infotainment displays.

All mentioned principles and elements do have effects on the resulting levels of situation awareness, end user benefit and acceptance and there is definitely need for research to assess the effect's quality and quantity. This will be done in future work.

8 Summary and Outlook

The current paper describes a systematic analysis to derive new ways of interaction between the driver and a partly automatic vehicle. Traffic jam situations are chosen as the example situational context for this paper since they are one of the first expected use cases for automated driver assistance systems.

In a partly automated driver assistance system, it is necessary that the driver is attentive and able to intervene at all times. On the other hand, it is supposed that one of the main benefits arising from automated driving, especially in traffic jams, is the ability to perform secondary tasks while being driven by the car. Hence, the aim of the activity is to find interaction schemes that are able to meet both requirements.

Two additional steps are included into the standard innovation process described by Pahl and Beitz [9]. In an Abstraction step the generalized functional principle is formulated for each good rated item found in the classic process. By doing that, an overview on the different ideas is gained. Four functional principles could be identified in the current analysis:

1. Artificial degradation of the systems performance (limiting the quality of control),
2. Artificial degradation of the system's availability (limiting the quantity of control),
3. Guidance of driver's attention towards the surrounding environment,
4. Monitoring the driver's attention towards the surrounding environment.

By formulating these generalized principles, a good overview on the resulting items can be given. In the subsequent Detailing step, the final results of the innovation process are derived only from these principles. Thus, the final results are decoupled from the previous results and new points of view are encouraged. Additional items that have not been thought of before arise. Figure 4 shows the quantity of items resulting from each step of the process. 20 % of the final results have not been mentioned before. This shows the efficiency of the given process.

A possible next step is to evaluate the resulting assistance systems. Further investigations of their effect onto the driver's level of situation awareness and to the driver's benefit and acceptance are necessary. Therefore the driver's interaction and response to such automated systems will be studied in future work.

References

1. Abendroth B, Bruder R (2009) Die Leistungsfähigkeit des Menschen für die Fahrzeugführung. In: Handbuch Fahrerassistenzsysteme. Grundlagen, Komponenten und Systeme für aktive Sicherheit und Komfort. Vieweg + Teubner, Wiesbaden
2. Bainbridge L (1983) Ironies of automation. Automatica 16(6):775–779
3. Endsley M (1995) Toward a theory of situation awareness in dynamic systems. Hum Factors 37(1):32–64
4. Freyer J et al (2010) Eine Spur aufmerksamer – Der Audi active lane assist. ATZ 112(12):926–930
5. Gasser TM et al (2012) Gemeinsamer Schlussbericht: BASt Projektgruppe „Rechtsfolgen zunehmender Fahrzeugautomatisierung". Bundesanstalt für Straßenwesen, Bergisch Gladbach
6. Gausemeier J, Ebbesmeyer P, Kallmeyer F (2001) Produktinnovation – Strategische Planung und Entwicklung der Produkte von morgen. Carl Hanser Verlag, München

7. Hörwick M, Wimmer M (2010) Fahrerüberwachungs- und Interaktionskonzept für hochautomatisierte Fahrerassistenzsysteme. Fortschrittsberichte VDI, Reihe 22: Mensch-Maschine-Systeme 32:123–135
8. Kaber D, Endsley M (1997) Out-of-the-loop performance problems and the use of intermediate levels of automation for improved control system functioning and safety. Process Saf Programs 16(3):126–131
9. Pahl G, Beitz W (2007) Konstruktionslehre – Grundlagen erfolgreicher Produktentwicklung – Methoden und Anwendung. Springer, Berlin
10. Rauch N et al (2009) The importance of driver state assessment within highly automated vehicles. In: Proceedings of the 16th world congress and exhibition on intelligent transport systems and services, pp 1–8
11. Rauch N (2009) Ein verhaltensbasiertes Messmodell zur Erfassung von Situationsbewusstsein im Fahrkontext. Universität Würzburg, Würzburg
12. Stanton N, Marsden P (1996) From fly-by-wire to drive-by-wire: safety implications of automation in vehicles. Saf Sci 24(1):35–49
13. Weilkes M, Buerkle L, Rentschler T (2006) Lane-keeping-support: from haptic lane-departure warning towards semi-autonomous lane guidance. In: Aachener Kolloquium für Fahrzeug- und Motorentechnik, pp 909–920
14. Winner H et al (2003) ACC adaptive cruise control. Robert Bosch GmbH, Plochingen

Integrated Simulation Toolset for ADA System Development

Martijn Tideman, Roy Bours, Hao Li, Tino Schulze and Takahito Nakano

Abstract This paper outlines the use of two complementary software packages that together provide the ability to cover all critical aspects of advanced driver assistance (ADA) system design. One of the software packages (PreScan) focuses on the design of the sensing and control systems, as well as the evaluation for a wide range of traffic and weather scenarios. The other software package (ASM) takes care of the detailed vehicle dynamics and vehicle control. By means of a real-world application example of a Lane Keeping Assistance system, the combined use of these software packages to design and evaluate ADA systems is demonstrated.

Keywords Active safety · ADA systems · Sensors · Vehicle dynamics · Simulation

---

F2012-I01-007

M. Tideman (✉) · H. Li
TASS, Shanghai, China

R. Bours
TASS, Hobart, The Netherlands

T. Schulze
dSPACE, Paderborn, Germany

T. Nakano
DENSO, Kariya, Japan

1 Introduction

The automotive industry is developing advanced driver assistance (ADA) systems and active safety systems that combine pre-crash sensing, driver warning (acoustic, haptic or visual) and vehicle dynamics control (braking, steering). The typical ADA system architecture is illustrated in Fig. 1.

Designers of ADA systems face many challenges. The involvement of different engineering disciplines—including sensing, electronics, control, vehicle dynamics, and safety—results in a complicated design process. Another challenge is that ADA systems need to function properly for the vast amount of different traffic scenarios which can occur on today's roads including effects of weather circumstances, road conditions and traffic density. This not only puts very stringent demands on system performance and reliability but also makes testing and verification complicated and time-consuming. Another complicating factor in the verification of ADA system designs is the need to test in reproducible conditions without the risk of endangering people. This factor often rules out field tests on public roads. Prototypes are expensive and unavailable early in the design process when changes can be made relatively easily and at low cost.

Virtual testing with simulation software provides an efficient and safe environment to design and evaluate ADA systems. Moreover, simulated scenarios are completely quantifiable, controllable and reproducible. Virtual testing is already commonly used in the design and evaluation of vehicle dynamics as well as crash safety systems. Recently, in the development process of ADA and active safety systems, simulation is becoming more popular as well. This paper describes how simulation software can be usefully applied within the design and evaluation of ADA systems.

In Sect. 2, general requirements for ADA system simulation software are described. Also, two commercially available software tools are introduced that jointly provide the required simulation capabilities to cover the critical aspects of ADA system design. Section 3 illustrates how these software packages can be combined to design and evaluate ADA systems. An application example for Lane Keeping Assistance is presented in Sect. 4. The results of this example are described in Sect. 5. Section 6, finally, contains concluding remarks.

2 ADAS and Active Safety Development

2.1 Traditional Testing Methods

Testing methods for passive safety systems are extensively defined and well-established. Regulations, such as FMVSS and ECE, and consumer programs, such as NCAP and IIHS, prescribe testing protocols in great detail and these testing methods are extensively applied in the development and certification of passive

Integrated Simulation Toolset for ADA System Development

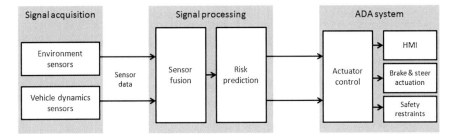

Fig. 1 Typical ADA system architecture

Fig. 2 Example of a pre-crash test in a test village [9]

safety equipment. Manufacturers have worked for many years on efficient development processes which include simulation, sled testing and full system crash testing.

In contrast with passive safety, testing protocols for active safety and ADA systems are hardly regulated and consumer protocols are not established. Only now, organizations as Euro NCAP are starting initiatives to award active safety systems. Euro NCAP has recently announced Euro NCAP Advanced, a reward system and standardized assessment procedure for new safety technologies. Unlike Euro NCAP's well established assessments involving physical tests at a crash laboratory, the new process is based entirely on the assessment of scientific evidence presented by the vehicle manufacturer and no test protocols are prescribed.

Several initiatives are working on developing standards describing system requirements and standard test programs. Some examples are the ISO standards [1], the Crash Avoidance Metrics Partnership initiative (CAMP) and NHTSA confirmation test requirements. In Europe, the Beyond NCAP group from EURO NCAP [2] and European projects such as PReVENT [3], Interactive and ASSESS [4] are working on standardization of test programs.

In recent years, the automotive industry has developed various testing procedures for ADA systems, ranging from test track protocols, pre-crash tests with stationary crash-forgiving foam dummy vehicles [5], plastic vehicle shapes attached to driving cars [6] and the use of vehicles that are driven by robots [7, 8]. Some manufacturers have even developed small test villages as shown in Fig. 2.

2.2 Challenges for ADA System Testing

The experimental methods for ADA system development have a number of drawbacks:

1. Experimental methods require the availability of system prototypes. Prototypes are expensive and unavailable early in the design process when changes can be made relatively easily and at low cost. Therefore, problems are found late in the development process and errors are expensive to fix.
2. Testing methods are time-consuming resulting in a limited coverage of the wide variety of traffic conditions.
3. The reproducibility of the test scenarios in test-track testing is limited, which makes it difficult to compare different systems in an objective way.
4. Many ground truth reference data cannot be measured (accurately) during test drives.
5. Critical scenarios are often too dangerous to test.

2.3 Virtual Testing of ADA Systems

Many of the drawbacks of hardware testing of ADA systems as described in the previous sections are not present for a virtual test environment. Virtual testing with simulation software provides an efficient and safe environment to design and evaluate ADA and active safety systems. Moreover, simulated scenarios are completely quantifiable, controllable and reproducible. Virtual testing is already commonly used in the design and evaluation of vehicles dynamics as well as crash safety systems. Recently, in the development process of ADA systems, simulation is becoming more popular. Virtual testing of ADA systems requires software that gives realistic representations of:

- Road objects (e.g. roads, lanes, sidewalks, curbs, light poles, traffic signs).
- Road users (e.g. pedestrians, bicyclists, cars, trucks).
- Environmental conditions (e.g. rain, snow, fog, lights, reflections, shadows).
- Sensing technology (e.g. camera, radar, lidar, infrared, ultrasonic, antennas, GPS).
- Vehicle dynamics (e.g. braking system, steering system, suspension).

Many perturbations of these different characteristics need to be simulated in order to assess the ADA system's reliability and robustness for the large range of conditions that can occur on real roads. This requires that the simulation models

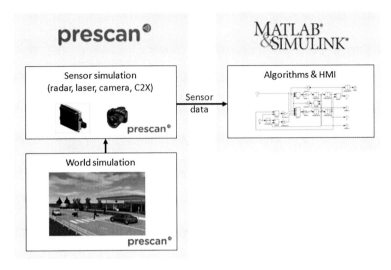

Fig. 3 Open-loop ADA system modeling using PreScan

are easy to create and modify. Also, fast simulation times are required to ensure that the vast of amount of simulations can be executed within the limited time available.

2.3.1 ADA System Modeling

PreScan is a physics-based simulation tool for the design and evaluation of intelligent vehicle systems, such as active safety systems, ADA systems and intelligent transportation systems [10].

The PreScan user interface allows users to build traffic scenarios using a database of road sections, infrastructural components (trees, buildings, traffic signs) and road users (cars, trucks, cyclists, pedestrians). Weather conditions (rain, snow, fog) and light circumstances can be modeled as well. The vehicle models can be equipped with models of different sensor types, including radar, laser, camera, ultrasonic, GPS and antennas for V2X communication. Sensor design and benchmarking is facilitated by easy exchange and modification of sensor type and sensor characteristics. An interface with MATLAB/Simulink enables users to design and verify algorithms for sensor data processing, sensor fusion, risk estimation, decision making and control (see Fig. 3).

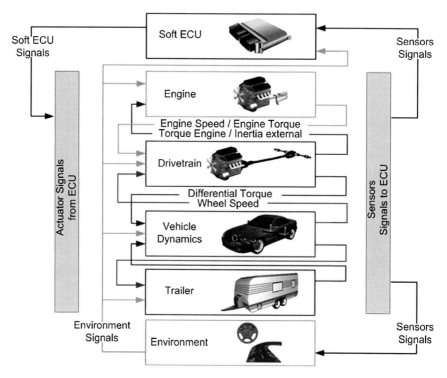

Fig. 4 Chassis system and control modeling using dSPACE automotive simulation models

2.3.2 Vehicle Dynamics Modeling

For ADA systems which control the vehicle's steering, braking or throttle actions (e.g. lane keeping assistance or emergency braking), a detailed model of the vehicle chassis and chassis controls are essential aspects of the system. dSPACE Automotive Simulation Models (ASM) is a simulation toolsuite offering a package for vehicle dynamics simulation. The package is an open Simulink model specially designed for Hardware-in-the-Loop (HIL) testing of ECUs—as well as for the controller design phase. Since the implementation of ASM Vehicle Dynamics, as depicted in Fig. 4, is an open Simulink model, additional components such as an interface to PreScan can be added easily.

The dynamics of the vehicle under test are described as a multi-body system which yields a set of differential equations of 2nd order. The necessary physical abstraction to model the relevant dynamics for ADA system function development and testing requires 10 degrees of freedom (DoFs), six for the body translation/rotation and four for the wheels. Furthermore the relevant kinematics and compliances and an appropriate tire modeling is necessary to simulate the dynamic behavior of the vehicle under test with an appropriate accuracy.

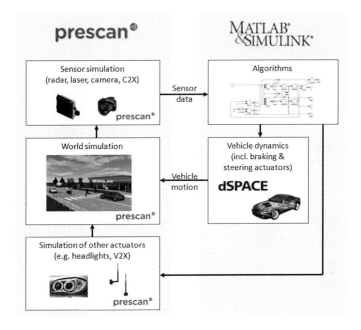

Fig. 5 Closed-loop ADA system modeling using PreScan and dSPACE/ASM software (overview)

3 PreScan: dSPACE/ASM Integration

Integration of both software packages makes it possible to investigate the robust functioning of closed-loop ADA system during complex vehicle maneuvers and detailed vehicle behavior.Therefore, a unique integration has been established between PreScan and dSPACE/ASM which allows closed-loop simulation of ADA systems with automated braking or steering actions. The vehicle motion is calculated in the dSPACE/ASM model. Updated vehicle positions and orientations are made available to the PreScan ego vehicle. Each time step, the world model is updated and new sensor signals are sent to MATLAB/Simulink for signal processing and actuation algorithm decisions (see Figs. 5, 6).

3.1 Driver Modeling

The ASM vehicle model is controlled by (1) a lateral controller resembling the steering behavior of the driver and (2) a longitudinal controller for the throttle and braking behavior of the driver.

The longitudinal controller (driver model for velocity control) is a Simulink block which translates a target velocity into throttle, clutch and braking signals for

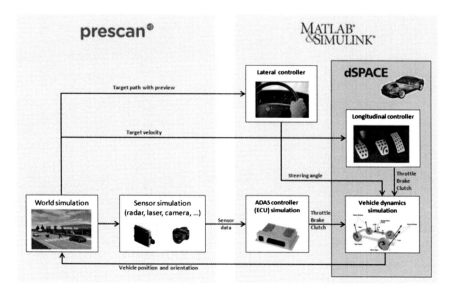

Fig. 6 Closed-loop ADA system modeling using PreScan and dSPACE/ASM software (detailed)

the vehicle model. The standard longitudinal controller in ASM is applied, and fed with a target velocity from the PreScan trajectory definition.

The lateral controller (driver model for steering actions) is a Simulink block which translates a trajectory into a steering angle input for the vehicle dynamics model. The standard lateral controller from PreScan is applied. The lateral controller uses a 10-point preview and is based on the optimal preview control theory as published by MacAdam [11, 12]. By introducing future path inputs, tracking models with preview information give much better tracking accuracy than 1-point compensation tracking model.

3.2 Frequency Subcycling

High-fidelity vehicle dynamics simulations are typically performed at 500–1000 Hz, whereas an ADA system typically runs at frequencies between 10 and 100 Hz. To facilitate this difference, PreScan and ASM are able to run at different frequencies. Within a typical ADAS setup, the *vehicle simulation* in ASM runs at 1000 Hz, whereas the *world simulation* and *sensor simulation* within PreScan run at 10–100 Hz. However, both the longitudinal and lateral *driver models* require input from PreScan with 500–1000 Hz. Therefore, PreScan still generates the *target path* and *target velocity* with a frequency of 500 Hz. This process of running the various PreScan modules at different frequencies is called "subcycling".

4 LKA Application

4.1 Introduction

By means of a real-world application example of a Lane Keeping Assistance (LKA) system, the combined use of these software packages to design and evaluate ADA systems is demonstrated.

4.2 LKA System Description

A Lane Keeping Assistance (LKA) system detects lane markers on the road, and assists the driver's steering to help keep the vehicle between lane markers. When the system detects the vehicle straying from its lane, it alerts the driver visually as well as with a buzzer, while applying a slight counter-steering torque, trying to prevent the vehicle from moving out of its lane. An LKA system consists of two key components: (1) a vision sensor, which detects shapes and positions of lane markers, and (2) a steering assist electronic control unit (ECU), which calculates a target steering torque based on the data from the vision sensor. The steering assist ECU then sends this steering torque signal to an electric power steering (EPS) ECU to control the EPS motor. Furthermore, the steering assist ECU determines if it is necessary to alert the driver about the vehicle's deviation from the lane. These alerts are communicated to the driver by a buzzer and a visual signal.

4.3 LKA Model

The simulation tool set as described in Sect. 3 was used to model a DENSO LKA system (see Fig. 7). Corresponding to the descriptions in Sect. 4.2, the LKA algorithm (which is modeled in Simulink) calculates the required steering torque based on the lane information derived from PreScan. Input signals used are, for example, "distance to line" and "road curvature". The steering torque is then fed into a steering model in ASM to prevent departure from the intended lane. Finally, arbitration between the lateral driver model and the LKA model is implemented to make sure that—after the LKA model is activated—the input from the lateral driver model is deactivated.

Simulation of ADA systems allows fast analysis of many different scenarios using automated parameter variations. Two important points had to be addressed for the LKA model:

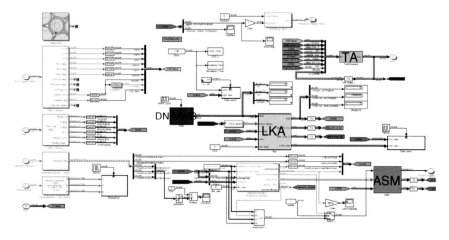

Fig. 7 Simulink representation of the LKA model including ASM vehicle dynamics, test automation (TA), LKA algorithm and PreScan sensor inputs

1. The performance of the lateral driver model to accurately control the ASM vehicle dynamics model. The PreScan lateral controller using a 10-point preview was optimized for the specific vehicle under test.
2. An efficient method for scenario generation. The lateral driver model was extended with an offset function allowing a constant or runtime-changeable lateral offset from the defined target path.

Due to these two modifications, it was possible to execute an automated test series of several hundred scenarios to verify the LKA performance.

5 Results

Using the LKA model as described in Sect. 4, the LKA system's basic functionality and performance is verified. Figure 8 depicts a scenario on a straight road, in which the driver is leaving the lane. The blue line shows the driver's steering input, whereas the red line shows the correction made by the LKA (thereby overruling the driver). It is shown that, in this scenario, the LKA works and that the vehicle gets back into its lane. It can also be seen that there is a slight deviation between the target path as input to the driver model (yellow line) and the actual path of the vehicle as controlled by the driver model and/or the LKA controller (purple line).

Integrated Simulation Toolset for ADA System Development 35

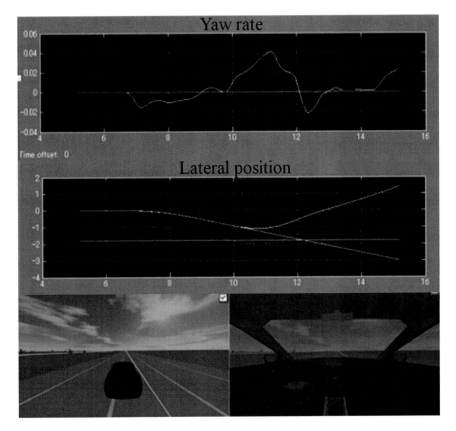

Fig. 8 Example results from the LKA simulation model

6 Conclusion

Advanced driving assistance (ADA) systems need to function properly in a vast amount of different traffic scenarios. This makes the development process of ADA systems a complicated, dangerous and time-consuming activity. Simulation software can help to overcome these difficulties by providing an efficient and safe environment for designing and evaluating ADA systems.

PreScan, dSPACE/ASM and MATLAB/Simulink form together a simulation tool chain that is suitable for virtual design and assessment of ADA systems such as ACC, CMB, PCS, LDW and LKA. This tool chain is capable of assessing the effects of the performance of all ADA system components (such as sensors, object detection & interpretation algorithms and vehicle dynamics control algorithms) on the safety performance of the complete ADA system. Unique in the methodology is the fact that different system-elements can be assessed using the same tool-set.

The use of the simulation tool chain was presented based on a real-world application example: DENSO's LKA system. It was described how the simulation tool chain is used to achieve a efficient and effective ADA system development process.

References

1. International Organization for Standardization website: www.iso.org
2. Euro NCAP website: www.euroncap.com
3. EC project PReVENT website: www.prevent-ip.org
4. EC project ASSESS website: www.assess-project.eu
5. Sala G (2003) Safety, legal issues, standards, Deliverable D5.01, European Commission, CHAMELEON project, Brussels, Belgium, January 28, 2003. Available at: www.crfproject-eu.org
6. Fecher N et al (2008) Test and evaluation methods for safety systems on the guidance level. Automatisierungstechnik 56:592–600
7. Gietelink O et al (2006) Development of advanced driver assistance systems with vehicle hardware-in-the-loop simulations. Int J Veh Syst Dyn 44(7):569–590
8. Schöner H-P et al (2009) Testing and verification of active safety systems with coordinated automated driving. In: Proceedings of the 21st international technical conference on the enhanced safety of vehicles ESV, Stuttgart, Germany
9. Autoliv website: www.autoliv.com
10. PreScan website: www.tass-safe.com/prescan
11. MacAdam CC (1980) An optimal preview control for linear systems. J Dyn Syst Meas Control. ASME 102(3). http://hdl.handle.net/2027.42/65011
12. MacAdam CC (1981) Application of an optimal preview control for simulation of closed-loop automobile driving. IEEE Trans Syst Man Cybern 11. http://hdl.handle.net/2027.42/65010

Haptic Velocity Guidance System by Accelerator Pedal Force Control for Enhancing Eco-Driving Performance

Feilong Yin, Ryuzo Hayashi, Raksincharoensak Pongsathorn and Nagai Masao

Abstract This paper describes a haptic velocity guidance system with the purpose of realizing better fuel efficiency and enhancing road capacity by using information from Intelligent Transport Systems (ITS). By controlling the accelerator pedal reaction force, the proposed system provides a desired pedal stroke and guides the driver to achieve a desired velocity in real time. The effectiveness of the proposed haptic velocity guidance method is verified by experiments using Driving Simulator (DS). The comparison with the other guidance HMI is presented via experimental results. The experimental results show that the proposed haptic guidance, which provides high accuracy and quick response in velocity tracking, is a promising velocity guidance HMI.

Keywords Pedal force control · Haptic velocity guidance · Human–machine-interface

1 Introduction

Nowadays, the automobile is absolutely essential throughout the world. On the other side, fuel consumption and automobile carbon dioxide emission have become serious energy and environmental problems, respectively. To solve these problems, eco-driving and eco-traffic have received considerable attention in recent years.

F2012-I01-012

F. Yin (✉) · R. Hayashi · R. Pongsathorn · N. Masao
Tokyo University of Agriculture and Technology, koganei, Japan

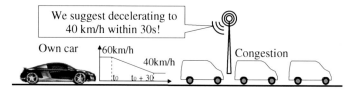

Fig. 1 Example of deceleration guidance application

Some devices based on eco-driving have been utilized in practice such as Eco-indicator of TOYOTA [1], Coaching and Teaching System of HONDA [2] and Eco-Pedal of NISSAN [3]. These systems reduce the consumption of fuel of each vehicle by restraining uneconomical acceleration, particular when the vehicle starts moving.

Another similar research is being developed by EU research project HAVEit. A bus applied the concept of "Active Green Driving" can use the right amount of energy according to the traffic environment and lower the fuel consumption with around 30 % compared with a traditional diesel engine [4].

However, this research focuses on eco-traffic guidance. As a future technology, management of traffic flows with ITS information is considered to realize the entire eco-traffic [5–8]. Figures 1 and 2 show the conceptual examples of the applications for such guidance. Figure 1 shows that the host vehicle saves fuel by starting deceleration gently ahead of traffic congestion out of sight. Figure 2 shows that a car is climbing up a slope. The driver may not notice the deceleration of own car which will cause traffic congestion. The velocity guidance system receives the warning and the suggestion of optimal speed from ITS infrastructure and guides the driver to recover the velocity quickly to avoid traffic jam. By conducting these, both fuel and time efficiencies can be improved significantly.

Traffic flow control requires highly accurate velocity guidance. Conventional velocity guidance methods include visual method such as indicators in instrument panel or on screen display of car navigation system and audio methods such as voice navigation by a speaker. In realization of accurate velocity guidance, visual methods have an advantage that it can show quantitative and continuous information as an indicator whereas an excessive gaze at the indicator may distract the driver's attention from ambient environment. On the other hand, frequent presentation of information may disturb the passengers' conversation or the enjoying of radio though audio methods do not distract the driver's attention from the environment so much. Since the haptic guidance has no such shortcomings, it is considered as the best way to accomplish velocity guidance.

In the aim of improving the fuel efficiency and traffic capacity, the authors propose a high accurate velocity guidance system. By changing the characteristics of the pedal reaction force, the proposed system informs the driver of the desired pedal stroke to realize an ideal velocity in real time. An experimental system is installed in the DS of Tokyo University of Agriculture and Technology (TUAT DS) (Fig. 3). The effectiveness of this velocity guidance system was presented in 22nd IAVSD.

Haptic Velocity Guidance System by Accelerator Pedal

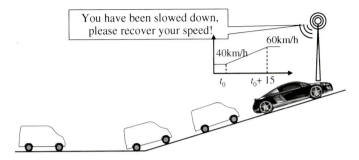

Fig. 2 Example of velocity recovering guidance application

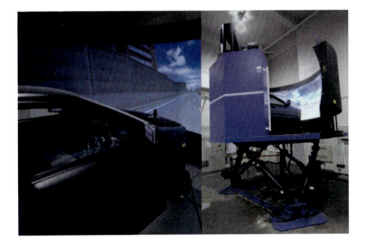

Fig. 3 TUAT driving simulator

2 Haptic Velocity Guidance System

The block diagram of haptic velocity guidance assistance system is shown in Fig. 4. This system has three basic functions: ITS information acquisition, desired pedal stroke calculation and desired pedal stroke guidance by haptic pedal.

3 Acquisition of its Information

The system gets ideal velocity profile from the ITS infrastructures for the guidance in the practical use. Since this research is carried out with a driving simulator and this paper intends to evaluate the proposed HMI, the accurate development of this function is not necessary. Accordingly, a desired (ideal) velocity profile based on typical driving is created to simulate this function.

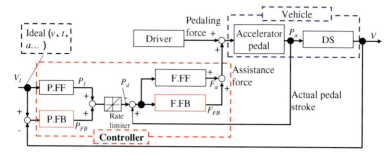

Fig. 4 Block diagram of haptic velocity guidance assistance system

4 Calculation of Desired Pedal Stroke

The desired pedal stroke (P_d) is calculated by two controllers: *Ideal Pedal Stroke Calculator* (P.FF) and *Velocity Error Feedback* (P.FB) controller.

The P.FF is a feed forward controller. P.FF is required to calculate the desired pedal stroke depending on the vehicle characteristics without considering the slope, wind, etc. However, a complex P.FF itself is a very tough research which needs a lot of time. In this paper, the ideal pedal stroke profile which realizes the ideal velocity profile is roughly calculated offline from the velocity response characteristic of the TUAT DS.

From previous experiments, it is found that the velocity error ($V_e = V - V_I$) mainly contains two parts, random error and steady state error that is caused by environmental variation, human delay or mistake, etc. Thus, a PI control described as follow is utilized in P.FB:

$$P_{FB} = \left(K_P + \frac{K_I}{s}\right) V_e \qquad (1)$$

Here P_{FB} is the adjusting pedal stroke, K_P is the proportional gain, K_I is the integral gain and s indicates the Laplace operator. Since the feed back of velocity is mainly related to the vehicle driving performance, the optimal parameters of the P.FB are observed from DS simulation. By comparing the accuracy of the actual velocity and the degree of fluctuation of the pedal stroke, the optimal set is: $K_P = 3$ %(km/h) and the $K_I = 0.1$ %/s/(km/h).

5 Desired pedal Stroke Guidance by Haptic Pedal

The guidance will be accomplished by changing the reaction force of the haptic pedal. To realize the characteristics control of the pedal reaction force, an additional pedal reaction force generation mechanism is installed in the DS. The pedal is connected to an AC servomotor via mechanical linkages. A low inertia motor is

Haptic Velocity Guidance System by Accelerator Pedal   41

Fig. 5 Haptic pedal force characteristics with respect to the pedal stroke deviation

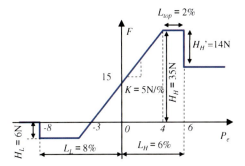

chosen and no gear transmission is used so that the additional mechanism does not obviously change the passive pedal characteristic. Additional reaction force with maximum value, i.e. 100 N could be generated by this mechanism.

The additional reaction force that guides the driver to desired pedal stroke is decided by two controllers: *Fixed Pedal Reaction Force Characteristic Controller* (F.FF), which is a feed forward controller, and *Pedal Stroke Error Feedback Controller* (F.FB), which is a feedback controller. The F.FF is designed according to the characteristics of passive reaction force and human stepping characteristics. The characteristic of F.FF is shown in Fig. 5 (the unit of pedal stroke is '%', varies from 0 to 100). Here horizontal axis indicates the pedal stroke error (P_e) which equals actual pedal stroke (P_a) minus desired pedal stroke (P_d), vertical axis indicates the additional reaction force of F.FF. There are two regions in additional reaction force: guidance region (pedal stroke error is from -8 to $+6$ %) and override region (pedal stroke error is over $+6$ % or under -8 %). The guidance region is designed to guide the driver to the desired pedal stroke by providing a click feeling when the driver passes the desired pedal stroke [9]. The balance of the stepping force, the passive pedal reaction force and additional reaction force is studied. Based on the study, the reaction force, where the actual pedal stroke equals to the desired pedal stroke, is designed to be 15 N and the largest additional reaction force is decided to be 35 N. From the previous experiment result, the drop force when the pedal stroke exceeds guidance region is decided to be -14 N.

F.FB is proportional to the integral of the pedal stroke error (P_e), and it is utilized to eliminate the steady state error of the pedal stroke.

$$F_{FB} = \frac{K_{P_e}}{s} P_e \quad (2)$$

An experiment to find the optimal K_{P_e} is carried out by three participants. By comparing the pedal stroke time history, the feedback of participants, K_{P_e} is decided to be 4 N/%.

The maximum guidance force that guides driver to release pedal is set to be 55 N and the maximum guidance force for guiding driver to step further is set to be -6 N. Such settings ensure that the guidance force could not operate the pedal itself.

Table 1 Participants information

NO.	Age	License history	Mileage/year (km)	Driving frequency
1	23	4	2,000	1/week
2	22	3	1,000	1/week
3	30	12	5,000	2,3/week
4	34	13	3,000	2,3/month
5	60	40	12,000	Everyday
6	62	43	9,000	4/week
7	23	3	100	4,5/year
8	23	3	0	0–1/year
9	33	14	800	1/month

6 Experiments for System Verification

An evaluation experiment to verify the accuracy of haptic velocity guidance by pedal reaction force control and compare with other types of guidance is carried out with TUAT DS. Questionnaires are done to learn the subjective feeling of the proposed system. The participants' information is shown in Table 1. Those are selected considering the variety of age and driving experience.

7 Guidance Technique

Four types of velocity guidance are used in this experiment: visual velocity guidance (V.visual), audio velocity guidance (V.audio), visual pedal stroke guidance (P.visual) and haptic pedal stroke guidance (P.haptic). A monitor is set in front of the instrument panel to achieve visual guidance (Fig. 6). The speaker of the TUAT DS is used to play the audio guidance and the haptic pedal mentioned above realizes the haptic guidance.

(a) *V.audio* The audio guidance contains the information of desired velocity and the time duration of adjustment. For example, if 'Please accelerate to 60 km/h in 15 s' is played and the actual velocity is 40 km/h, the driver should do his best to accelerate from 40 to 60 km/h with a constant acceleration around 0.37 m/s^2.

(b) *V.visual* (subgraph (b) of Fig. 6) The green bar indicates the actual velocity and the yellow bar indicates the current desired (ideal) velocity (subgraph (a) of Fig. 5). During the guidance, the driver is instructed to follow the yellow bar as well as possible.

(c) *P.visual* (subgraph (c) of Fig. 6) The green bar indicates the actual pedal stroke; the yellow bar indicates the desired pedal stroke (as shown in Fig. 4) including velocity error feedback.

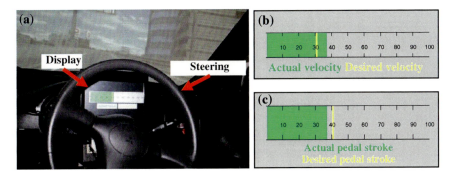

Fig. 6 Image of visual device

(d) *P.haptic* The haptic guidance guides the driver to the desired pedal stroke only by controlling the pedal reaction force. The meaning of the reaction force variation is explained to the participants as follow:

The force rises: the participant should further release the pedal.
The force decreases: the participants should further press the pedal.

8 Experiment Scenario

In order to simulate the real driving situation, Metropolitan Expressway of Tokyo which has many curves is selected to carry out this experiment. The primary task is to keep the lane as usual. The task performance will be examined by checking the maximum value of the lateral displacement. Here lateral displacement is the distance from the host car gravity center to the lane's center. The secondary task is to follow the guidance (velocity or pedal stroke). Training is executed before the experiment to make sure that every participant is familiar with the guidance methods. The ideal velocity profiles used in training and experiment are different.

9 Experimental Results

9.1 Lateral Displacement Comparison

The maximum value of lateral displacement is shown in Fig. 7, where the vertical axis indicates the displacement (m), horizontal axis indicates the guidance. The lines with different colors and marks indicate different drivers. The variety of displacements of every participant (except driver 2 and 7) is not obvious, almost in

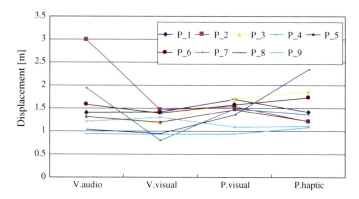

Fig. 7 Maximum value of the lateral displacement

the region of 1–2 m, which means that the participants accomplished the primary task very well. The ranges of the values are different from participant to participant caused by individual difference. The lateral displacement data shows that the result of this experiment is credible.

9.2 Velocity Comparison

As an example, velocity time histories of participant 3 are shown in Fig. 8. Here (a) is the result of audio velocity guidance (b) is the result of visual velocity guidance (c) is the result of visual pedal stroke guidance and (d) is the result of haptic pedal stroke guidance. The vertical axis indicates the velocity (km/h), horizontal axis indicates the time, the blue lines indicate desired velocity and the red lines indicate the actual velocity. The actual velocity of haptic guidance shows better accuracy, smoothness in following the desired value and the reaction time is shorter than visual guidance and audio guidance as well.

Instead of showing all the time history of all the participants, the RMS of velocity error of all experiments is shown in Fig. 9, where the vertical axis indicates the RMS (km/h), horizontal axis indicates the guidance. For all participants, the RMS of velocity error of haptic guidance is the smallest and nearly the same, which means haptic guidance suits this task very well.

9.3 Pedal Stroke Comparison

Pedal stroke guidance time histories of participant 3 are shown in Fig. 10. Here (c) is the result of visual pedal stroke guidance and (d) is the result of haptic pedal

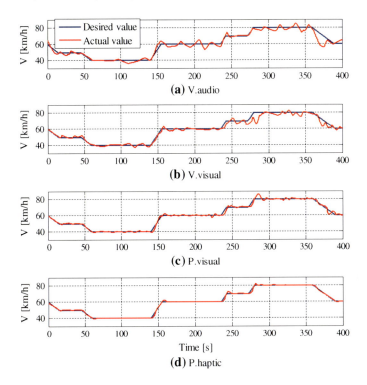

Fig. 8 Velocity time history of participant 3

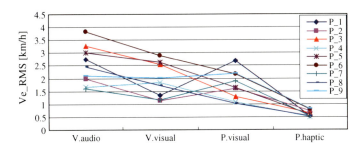

Fig. 9 RMS of velocity error of all guidance and all participants

stroke guidance. The vertical axis indicates the pedal stroke (%), horizontal axis indicates time; the blue lines indicate desired pedal stroke and the red lines indicate the actual pedal stroke. The desired pedal stroke of P.visual and P.haptic is the sum of ideal pedal stroke and adjusting pedal stroke. The result of haptic guidance is better than the result of visual pedal stroke guidance; it shows that the haptic guidance is easier to follow. The pedal stroke of haptic velocity guidance is smoother, which infers higher fuel efficiency (Fig. 11).

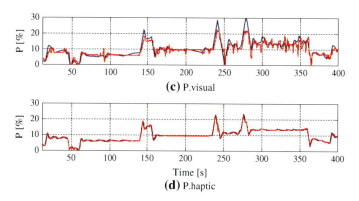

Fig. 10 Pedal stroke time history of participant 3

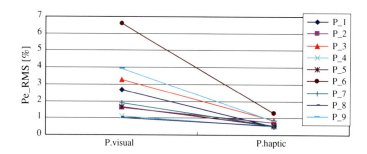

Fig. 11 RMS of pedal stroke error of pedal stroke guidance and all participants

During velocity guidance, the drivers are not required to follow the desired pedal stroke. Therefore only the RMS of pedal stroke error of *P.visual* and *P.haptic* is shown in Fig. 8. It can be seen that the accuracy of *P.haptic* of all participants are almost the same, and smaller than *P.visual*. It can be observed that the effect of haptic device is superior comparing with visual device from this experiment.

9.4 Pedal Stroke Tracking

Another experiment is arranged to study the driver's response during haptic guidance. An example of pedal stroke guidance is shown in Fig. 12. Subfigure (a) shows time history of desired pedal stroke P_d (in black) and time history of actual pedal stroke P_a (in blue). Subfigure (b) shows time history of guidance force F_a. The red line is the pedal stroke calculated by the driver-pedal model which will be described latter. The followings could be observed:

(a) The guidance force normally keeps constant while the desired pedal stroke is invariant.

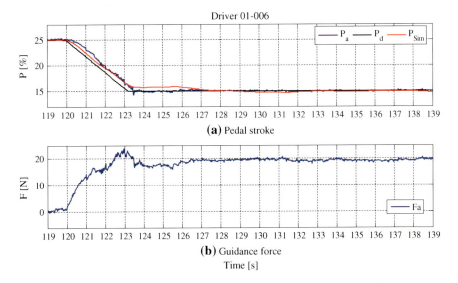

Fig. 12 Time history of pedal stroke guidance

(b) It can be noticed that drivers follow the desired pedal stroke with a little delay. Due to this the guidance force changed quickly right after the desired pedal stroke changed.
(c) The actual pedal stroke normally decreases when the guidance force increases (from 120 to 123 s). However, the actual pedal stroke keeps decreasing even the guidance force decreases after a period of increase (from 123 to 124 s).

Above all, it can be concluded that the pedal operation is related to the change of guidance force. In this paper, a model as expressed in Eq. 3 is employed to modify driver-pedal system. Here, driver is supposed to control pedal according to three values, the differential of guidance force ($\triangle F_a$), the integral of $\triangle F_a$ and the changing rate of $\triangle F_a$. Here, $\triangle F_a$ equals to recent guidance force deducts guidance force before desired pedal stroke changed. The gains are K_P, K_I and K_D, respectively. This model is called TPID model in this paper. The pedal stroke in time domain is described by Eq. 4. Here, $P_a(0)$ is the pedal stroke before the variation of desired pedal stroke and $F_a(0)$ is the guidance force before the start of desired pedal stroke changing.

$$G(s) = \left(K_P + K_I \frac{1}{s} + K_D s\right) e^{-\tau_L s} \qquad (3)$$

$$P_a(t) = P_a(0) + K_P \Delta F_a(t - \tau_L) + \int_0^t K_I \Delta F_a(t - \tau_L) dt + K_D \frac{d\Delta F_a(t - \tau_L)}{dt},$$

$$\Delta F_a(t) = F_a(t) - F_a(0) \qquad (4)$$

The parameters are identified by using least squares method. The data of guidance force is processed by a low pass filter and the threshold frequency is 5 Hz before calculating the rate of guidance force. The identified equation of Fig. 12 is:

$$G(s) = \left(-0.515 - 0.0014\frac{1}{s} + 0.00074s\right) e^{-1.081s} \tag{5}$$

The pedal stroke calculated by this equation is the red line in the Fig. 12a. From figure, it can be found that the simulated pedal stroke matched the actual pedal stroke very well and the mean absolute error of simulated pedal stroke is only 0.37 %. For most experimental data, there is a set of parameters that can give an accurate pedal stroke simulation result. Moreover, the K_I and K_D are always very small. This means that pedal stroke follows the haptic guidance mainly according to $\triangle F_a$ while time delay exists.

10 Conclusion

This paper proposes velocity guidance system by haptic interface-a haptic accelerator pedal. This system intends to realize better traffic efficiency by guiding the driver to drive with proper velocity. A haptic pedal is designed and the haptic guidance maneuver is discussed while the other functions are neglected or simply developed.

Four types of HMI are applied in the comparison experiments carried out with TUAT DS. The experimental results show that the drivers finished the lane keeping task with nearly the same accuracy. The haptic guidance method realized higher accurate velocity guidance than the other three. However, the comparison is not sufficient. In future research, comparison with other types of velocity assistance system, e.g. ACC, should be carried out as well.

A model is proposed that can accurately simulate the response of driver-pedal system. Since the friction exists in accelerator pedal, the modelling of pedal is complex. Moreover, human driver is in the loop, it is difficult to express haptic guidance only by using a simple model as proposed. The effort to find a more suitable model should be done in future research.

References

1. Toyota Motor Corporation, News Release, Toyota to introduce eco drive indicator. Available http://www.toyota.co.jp/en/news/06/0929.html
2. Fujiki Y, Takemoto H, Enomoto K (2009) Ecological drive assist system. J Soc Automotive Eng Japan 63(12):88–91 (in Japanese)
3. Sakaguchi S, Shiomi M, Oomori M (2010) Development of ECO pedal system assisting eco-driving. JSAE proceedings (Society of Automotive Engineers of Japan) 48(10):1–6
4. Highly Automated Vehicles for Intelligent Transport, Predicting the future for advanced energy management. Available http://planetgreen.discovery.com/tech-transport/eco-driving-save-fuel.html

5. EcoDriving USA, EcoDriving practices. Available. http://www.ecodrivingusa.com/#/ecodriving-practices/
6. Kestinga A, Treibera M, Schönhofa M, Helbing D (2008) Adaptive cruise control design for active congestion avoidance. Transport Res C: Emerg Tech 16(6):668–683
7. Arnott R, Small K The economics of traffic congestion. Am Sci 82:446–455
8. Alvarez L, Horowitz R, Li P (1999) Traffic flow control in automated highway systems. Control Eng Pract 7:1071–1078
9. Nagurka M, Marklin R (1999) Measurement of impedance characteristics of computer keyboard keys. In: Proceedings of the 7th Mediterranean conference on control and automation (MED99), p 1940–1949

Physical Model-Based Yaw Rate and Steering Wheel Angle Offset Compensation

Stefan Solyom and Johan Hultén

Abstract The article describes a model-based estimation method for offsets in the yaw rate measurement and the steering wheel angle measurement for a vehicle. The estimator is constructed as a hybrid system comprising of two linear estimators. The switching between the two local linear systems is based on the vehicle operating mode.

Keywords Estimation · Linear regression · Bicycle model

1 Introduction

Yaw rate and steering wheel angle are two basic informations in a vehicle that are used, among others for lateral stability control. These sensor signals are known to be subject for offsets that in some cases can even be time varying.

The typical offset compensation algorithms implemented in production vehicles are based on averaging the steering wheel angle at high speeds where a straight ahead position is likely to occur, and comparing the measurement yaw-rate and computed yaw rate from wheel speeds for the steering wheel offset and yaw rate offset respectively [1]. These algorithms are rather robust; however the precision and the convergence speed might be rather low. In particular, a typical

---

F2012-I01-014

S. Solyom (✉)
Active Safety and Vehicle Dynamics, Volvo Car Corporation, Gothenburg, Sweden

J. Hultén
Sentient+, Gothenburg, Sweden

performance for the steering wheel angle offset is a lower bound on the detectable offset of some degrees, i.e. the straight ahead position is known with an error of some few degrees of the steering wheel angle. The next generation vehicles are equipped with electric power assisted steering (EPAS), which are enablers for sophisticated features such as emergency steering assistance in case of imminent accidents, automatic parking and sophisticated steering feel algorithms. These algorithms have much higher requirements on e.g. straight ahead information, that is the offset on the steering wheel angle information. In some cases it is desirable an improvement in the scale of an order of magnitude, in comparison to the todays available offset information.

There are various approaches in the literature, from the direct estimation of yaw rate and steering wheel angle offset [2], to Kalman filter based approach for yaw rate offset estimate [3]. The method proposed in this article treats both steering wheel angle offset and yaw rate offset in the same time. Such an approach is necessary when using a bicycle model, since the two entities (yaw rate and steering wheel angle) will be coupled through the bicycle model. It is also shown that in order to get a high precision estimate of the steering wheel angle offset it is necessary to estimate vehicle parameters as well.

2 The Bicycle Model

The algorithm is constructed to estimate the offsets during normal, everyday driving. This means that a linear bicycle model would suffice for describing the vehicle behavior. Moreover, in order to reduce the influence of model uncertainties such as road conditions, an additional stationarity condition is imposed. That is, the algorithm is activated only during stationary linear driving conditions. It is worth mentioning that driving conditions around the straight ahead position will most of the time meet the linear and stationarity requirements.

From [4], imposing stationarity conditions on the bicycle model, the yaw gain is given by:

$$k_r = \frac{r}{\delta} = \frac{(a+b)v}{(a+b)^2 + \kappa v^2} \tag{1}$$

with,

$$\kappa = m\left(\frac{b}{C_f} - \frac{a}{C_r}\right), \tag{2}$$

and:

- a, b are the distance from the COG to the front and rear axles, respectively,
- m is the mass of the vehicle,
- C_f, C_r are the cornering stiffnesses of the front and rear axles respectively,

- v is the longitudinal speed of the vehicle,
- r is the yaw rate of the vehicle,
- δ is the road wheel angle of the vehicle. This is a scaled version of the steering wheel angle.

The entity κ, is a scaled value of the understeer coefficient. The scaling factors are the wheel base (i.e. $a + b$) and the gravitational constant, which are considered constant.

3 The Linear Regressions

Then from Eq. (1), considering an offset on the yaw rate and the steering wheel angle one can write:

$$r + \Delta r = k_r(\delta + \Delta\delta). \qquad (3)$$

Considering a sensor measurement on the yaw rate r_m and the steering wheel angle, translated to the road wheel angle δ_m, the following linear regression for the offsets can be formed:

$$r_m - k_r\delta = \begin{bmatrix} 1 & k_r \end{bmatrix}^T [\Delta r \quad \Delta\delta], \qquad (4)$$

where the parameters to be estimated are the yaw rate and road wheel angle offset and the regressor is basically formed by the yaw gain k_r.

3.1 Yaw Gain Variation

The yaw gain is not a measurable entity, but it is computable, as shown in (1) from the wheel base, the vehicle speed and κ. The wheel base is constant and known, the vehicle speed is measurable but κ needs to be computed. Unfortunately this entity can vary from one driving cycle to another through the vehicle loading m, but also within a driving cycle due to the road conditions, i.e. the cornering stiffnesses C_f, C_r. Figure 1 shows a graphical representation of the possible variations of the yaw gain due to varying κ. Also the nonlinear dependence on the vehicle speed is shown. Two extreme values of kappa are considered by taking a cf, cr/cf combination within the polytope (70000, 120000) × (1, 1.4). The higher pair of curves (red and blue) is given by a nominal loading of the vehicle and a 10 % error in the COG position in the direction of the rear axle. The lower pair of curves show a vehicle with 40 % increased loading and a 10 % error in the COG from its nominal value, in the direction of the front axle. It is transparent now that the yaw gain presents a significant variation with κ. However, perhaps the most relevant question is how much error this variation would induce on the estimated offsets.

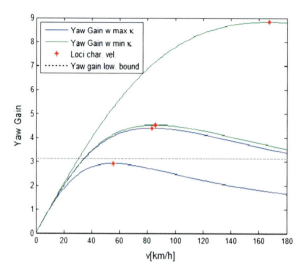

Fig. 1 Yaw gain dependence on vehicle speed. The vehicle parameters (included in kappa). Two extreme values of kappa are considered by taking a cf, cr/cf combination within the polytope (70000,120000) × (1,1.4). The higher pair of curves (*red* and *blue*) is given by a nominal loading of the vehicle and a 10 % error in the COG position in the direction of the rear axle. The lower pair of curves show a vehicle with 40 % increased loading and a 10 % error in the COG from its nominal value, in the direction of the front axle

3.1.1 κ-Dependence

Since todays and next generation EPAS applications the steering wheel angle precision requirements are very severe, it is of high interest to analyze the influence of κ variations on the steering wheel angle offset. Consider a given yaw rate r and assume that there is no offset on the yaw rate. This assumption is somewhat a worst case scenario, because this way the entire uncertainty will propagate to the road wheel angle offset. Moreover consider an uncertainty on κ act on the system, with the two extreme values $\kappa_{min, max}$, then the worst case stationary error on the steering wheel offset becomes:

$$\Delta\Delta\delta_{swa} = \frac{r_{max}\tau}{(a+b)}|k_{min} - k_{max}|v_x. \qquad (5)$$

In terms of the yaw gain (5) is equivalent to:

$$\Delta\Delta\delta_{swa} = r_{max}\tau\left(\frac{1}{k_{rmin|v_x}} - \frac{1}{k_{rmax|v_x}}\right). \qquad (6)$$

With a road wheel to steering wheel ratio of around 12, the κ variation for the same loading in the two different loading cases considered in Fig. 1, rmax = 3°/s and vx = 90 km/h, this error becomes around 3° and 6° on the steering wheel,

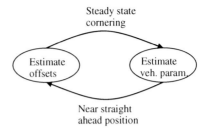

Fig. 2 The estimator switches between two local estimators depending on the driving situation

respectively. It is interesting to note that the κ variation for each pair of curves is around 300 % (a typical wheel base can be considered to be 2.5 m).

It can be argued that during normal driving there will be an averaging effect around zero angles which will mean that these errors will not get through completely the filtering effect of the estimation algorithm. However, it is highly unlikely that such an effect can be guaranteed to such extent that a high precision of the offset estimates can be guaranteed nor that the convergence speed of the algorithms will be satisfactory. Since the magnitude of the errors from the analysis above is not acceptable, it is clear now that a fixed nominal value for κ can not be used, but it needs online estimation.

3.2 κ Estimation

The online estimation of κ can be done with a linear regression from relation (1), i.e.:

$$(a+b)[r(a+b) - \delta v_x] = \kappa v_x^2 r \qquad (7)$$

where the parameter to be estimated is κ, the regressor is $v_x^2 r$, and the measurement is $(a+b)[r(a+b) - \delta v_x]$.

4 The Estimation Algorithms

The estimation algorithm consists of two parts, one estimating the steering wheel angle and yaw rate offset using (4), while the other is estimating the vehicle parameter κ from (7). These two algorithms are to complement each other, since while the first is to operate at a near straight ahead position, the other is to function at steady state cornering situations, as shown in Fig. 2. The switching between the two local estimators is realized based on some vehicle states such as lateral acceleration, yaw rate, steering wheel angle and speed and vehicle speed.

The typical driving situations where κ is estimated, are roundabout driving, where the steering wheel angle is often above 100°. In this case the existing offset on

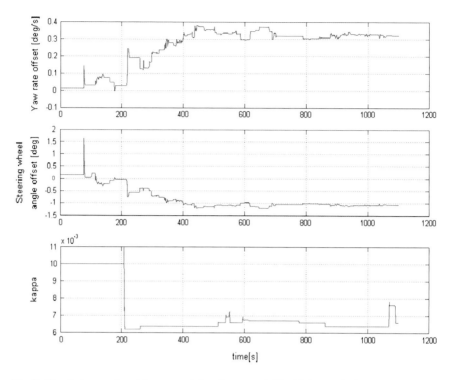

Fig. 3 Simulation results with measured data from a real-life driving cycle. The yaw rate offset is estimated with a steady state error less than 0.05 °/s. The steering wheel angle offset is estimated with a steady state error less than 0.1 °. κ is estimated with a steady state error of less than 20 %

the steering wheel and yaw rate is neglected, however small values of these offsets will not make a big impact on the κ estimate. The expected estimation error on κ is of some percent, for an error of a couple of degrees on the steering wheel angle. This way, convergence of both offset estimates and κ estimates is to be expected.

Both the yaw-rate and steering wheel angle offset are estimated in the same time as they form a linear regression. The typical driving situations where the offsets can be estimated are at steady state linear driving near straight ahead position, which is considered to be with the steering wheel angle up to around 50°. The excitation in the regressors is given by the variation in the yaw gain, obtain by varying the longitudinal speed of the vehicle.

4.1 Persistence of Excitation

For the estimate of the steering wheel angle and yaw rate offset, the excitation is basically delivered by the vehicle speed through the yaw gain variation. Since the

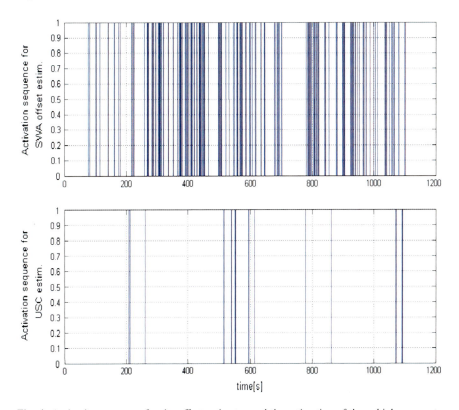

Fig. 4 Activation sequence for the offset estimates and the estimation of the vehicle parameter

yaw gain is a nonlinear function of the velocity, as shown in Fig. 1, at the top of the yaw gain characteristics a velocity interval will give a small change in the yaw gain, i.e. there is loss of persistence of excitation. In practice, due to noise, this means that a velocity interval around the top of the yaw gain curve (this point is given by the characteristic velocity of the vehicle), will not give any new information for the estimator. In this interval the estimator is to be shut down. This can be automatically done by using weights dependent on the values of k_r.

5 Simulation Results

The local estimators used in this experiment are linear weighted recursive least squares [5].

The algorithm have been tested against a real-life measured driving cycle, where among others, the vehicle speed and the steering wheel angle were logged. These signals where then fed into a bicycle model. The reason for this is to have control on the vehicle parameters, i.e. κ. Figures 3 and 4 show the simulation results, where an artificial yaw rate offset was introduced in the measurement, of 0.3°/s, a steering wheel offset of 1° and a system with a true κ of 0.0054 is used. The yaw rate offset is estimated with a steady state error less than 0.05°/s. The steering wheel angle offset is estimated with a steady state error less than 0.1°. κ is estimated with a steady state error of less than 20 %.

6 Conclusions

A hybrid estimation scheme is presented for high precision steering wheel angle and yaw rate offset estimates. The superior precision the offset estimates is possible due to the online estimation of the vehicle parameter κ.

Simulation results are presented using real-life driving cycle data with satisfactory results, where the yaw rate offset is estimated with a steady state error less than 0.05°/s. The steering wheel angle offset is estimated with a steady state error less than 0.1°.

References

1. Williams DA (1991) Apparatus for measuring the yaw rate of a vehicle. US Patent Application nr. US5274576. Group Lotus PLC
2. Zhenhai G (2003) Soft sensor application in vehicle yaw rate measurement based on Kalman filter and vehicle dynamics. Proc Intell Transp Syst 2:1352–1354
3. Gustafsson F et al (2001) Sensor fusion for accurate computation of yaw rate and absolute velocity, SAE technical paper. SAE, Warrendale
4. Pacejka H (2006) Tyre and vehicle dynamics, vol Second edn. Elsevier, Oxford
5. Åström KJ, Wittenmark B (1997) Computer-controlled systems. Prentice Hall, Englewood Cliffs

Dominant Driving Operations in Curve Sections Differentiating Skilled and Unskilled Drivers

Shuguang Li, Shigeyuki Yamabe, Yoichi Sato,
Takayuki Hirasawa, Suda Yoshihiro, P. N. Chandrasiri,
Kazunari Nawa, Takeshi Matsumura and Koji Taguchi

Abstract Our objective is to develop a new driving assist system that can help low-skilled drivers improve their driving skill. In this paper, we describe a statistical method we have developed to extract distinctions between high- and low-skilled drivers. There are three key contributions. The first is the introduction of wavelet transform to analyze the frequency character of driver operations. The second is a feature extraction technology based on AdaBoost, which selects a small number of critical operation features between high- and low-skilled drivers. The third is a simple definition for high- and low-skilled drivers. We performed a series of experiments using a driving simulator on a specially designed course including several curves and then used the proposed method to extract driving operation features showing the difference between the two groups.

Keywords Driver behavior · Driving simulator · Driving skill · Features extraction · Curve sections

F2012-I01-016

S. Li (✉) · S. Yamabe · Y. Sato · T. Hirasawa · S. Yoshihiro
The University of Tokyo, Tokyo, Japan

P. N. Chandrasiri · K. Nawa
Toyota Info Technology Center Co. LTD, Tokyo, Japan

T. Matsumura · K. Taguchi
Toyota Motor Corporation, Toyota, Japan

1 Introduction

One way to improve driving skill is to supply effective advice to low-skilled drivers. Drivers' skills determine how safely they drive, how much fuel they consume, and even how comfortable they feel while driving. To advise low-skilled drivers effectively, it is important to understand the different characteristics of high- and low-skilled drivers. Previous studies suggest that high-skilled drivers share certain characteristics in accelerator, brake, and steering operations [1-3]. Here, we propose a statistical method to extract these characteristics.

The proposed technology extracts critical features between high- and low-skilled drivers from a huge database obtained via telematics. Using telematics, we can organize floating car data in an information centre, and after data processing, traffic information is sent to terminal units such as a car navigator or even smartphones. The spread of such systems creates a huge database in the information centre, which the proposed method can access to extract the driving features shared by high-skilled drivers.

Many researches have been focused on drivers' skill in recent years. For example, researchers have introduced jerk, which is the derivative of acceleration with respect to time, as a criterion to determine a driver's status. Murphy et al. showed that less fuel is consumed and safety is improved if a vehicle moves smoothly, which means a relatively smaller jerk [1]. Motivated by their research, jerk and average speed are adopted as an effective means of defining a drivers' skill in this study. Other researchers have also tried to use driving skill to adapt vehicle control parameters to facilitate a specific driver's needs in terms of vehicle performance and safety [2]. They found out that the Discrete Fourier transform (DFT) coefficients of the steering wheel angle were very different depending on the skill levels of drivers and then used these coefficients as discriminant features to construct a driving skill recognizer via pattern recognition algorithms. Their paper suggested that the frequency characteristics of the drivers' operations were very different due to driving skill. However, the DFT coefficients merely reflect the average global characteristics, and the effect of the locations at which the operational features appeared was not considered. We therefore use wavelet transform in the proposed method so that we can obtain different frequency components of driving operations corresponding to locations. The extracted results can then show where and how the skilled drivers' operation characteristics tend to appear. This is key in terms of supplying effective advice to low-skilled drivers.

In this paper, we extracted a small number of relevant operational features using AdaBoost [4]. Feature extraction has long been studied and many statistical methods for feature extraction have been proposed that are now widely used in communication, signal and image processing, face detection, speech cognition, character recognition, and so on [5-7]. Perhaps the best known of these methods are principal component analysis (PCA) and neural network (NN). In both PCA and NN, the dimensional pattern of the output must be selected and presented to the network to test whether it can reach a target criterion that regularly aims to minimize an error function, such as the mean square deviation. If we had enough previous knowledge, the number of candidate pattern combinations would be relatively small and we would be able to obtain quick results. However, our objective in this study is to extract the driving features between high- and low-skilled drivers, and previous knowledge is not sufficient to determine the dimension of output. Even if it were, the number of candidate feature combinations would be so high that it would be extremely difficult to decide which combinations would be the most critical features between groups. To overcome this problem (and motivated by [4]), we decided to select features via AdaBoost. In the AdaBoost algorithm, a combination of candidate features is not necessary, as each weak classifier depends on only a single feature. As a result, only one feature can be selected at each step of the boosting process, and at the end of each stage, all selected weak classifiers are combined to form a final strong classifier using a weighted majority vote. This strong classifier is used to test the training data, and when the correct classification rate reaches a pre-set value, the feature extraction process is finished. A few classifiers are thus created, each of which contain an objective feature.

In the proposed method, the strategy is to create a database, prepare candidate features, and finally extract features. To obtain the necessary data, we performed a series of experiments using a driving simulator and established a feature extraction method based on the collected database. First, high- and low-skilled groups were defined as training data by average speed and composed jerk. Next, a large number of candidate features derived from driving operations were prepared and input into an AdaBoost algorithm. Finally, a small number of critical features to differentiate the two groups were automatically extracted. We describe the proposed method and the experimental results in detail in the following sections.

2 Driving Feature Extraction Algorithm

The flowchart of the proposed feature extraction method is shown in Fig. 1. First, we obtained driving data by experiments using a driving simulator. Next, we added wavelet transform to analyze the frequency characteristics of the driving operations. In principle, high-skilled drivers drive smoothly while low-skilled ones are more wobbly. In other words, their operating frequency is different even at the same location. Therefore, to compare the operation characteristics between laps, driving operation data including steering angle, accelerator, and brake pedal position, along with the corresponding operating speed, was decomposed into different frequency components by wavelet transform. However, it was still difficult to compare between laps using time history data because the average speed of each lap is different. We therefore normalized all decomposed data by distance so that we could compare operation at the same location between laps. In the next step, a large quantity of candidate features were produced by adding a sub-windows function to the most relevant frequency components so that the

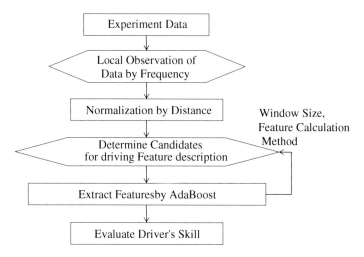

Fig. 1 Proposed feature extraction algorithm

candidate features could reflect variations of drivers' operations in a short period. Finally, we extracted a few driving operation features between high- and low-skilled drivers.

3 Experiment

3.1 Outline of Experiments

All experiments were performed using a driving simulator equipped with a 6-DOF motion platform (Fig. 2). The three screens and projectors offered drivers a 140° field-of-view. Many other experiments on driving behavior have been performed using this DS.

In this paper, our focus is driving skill in an urban environment with curves. All experiments were performed using the same course and scenario (Fig. 2b, c). The course included six left-hand curves with different radii, each of which was separated by a straight 300-m section. To ensure that all drivers entered the curves at almost the same speed, two speed limit signs were placed 50 and 100 m ahead of the start point of each curve (Fig. 2c). Drivers had to adjust their speed by about 60 km/h within this interval. They were permitted to drive freely outside of the interval, with the only condition being that they avoid veering off the lane. Each driver completed 12 laps, the first two of which were practice laps. Sixteen drivers participated in the experiment.

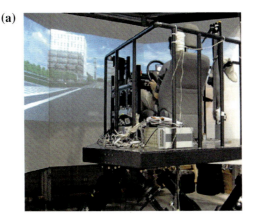

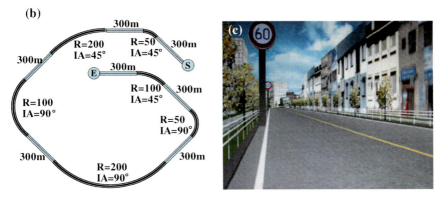

Fig. 2 (**a**) Driving simulator. (**b**) Experiment course. (**c**) Experiment scenario

3.2 Driving Skill Level Definition

Driving skill definition is not a simple issue. Most researchers use driving experience as criterion, and of course, a different definition can arise from a different viewpoint. In this study, we chose to use multiple indicators. In principle, indicators such as yaw rate, steering angle, velocity and jerk, etc. can be used to define driving skill. For example, a high-skilled driver can finish a lap with a high speed, small yaw rate variation, and even less fuel consumption. As the number of indicators increases, the definition becomes stricter and stricter, and the dimension of high-skilled driving groups descends. If the training database is large enough, more indicators can be adopted.

In the data pre-processing, we defined high- and low-skilled groups using a summation of the composed jerk and average. As mentioned above, researchers have found that less fuel is consumed and safety is improved if a vehicle is driven

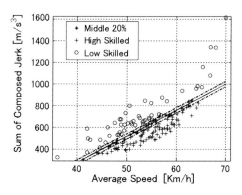

Fig. 3 Definition of high- and low-skilled driving

smoothly, which means a relatively smaller jerk [1]. Here, we introduce the parameter J, which is the integrated value of the composed jerk:

$$J = \sum_{i=1}^{N} \sqrt{J_{lateral-i}^2 + J_{longitude-i}^2}, \quad (1)$$

where $J_{lateral-i}$ and $J_{longitude-i}$ function as the *i-th* sample of normalized lateral and longitude jerk, respectively, and N is the total number of the normalized jerk samples from the start point to the end.

The basic concept of our definition is that if two laps have an equal average speed, the lap with the smaller J is selected as the more highly skilled. As an example, a definition for curve 1 is shown in Fig. 3, where the x-axis is the average speed and the y-axis is J. We used a regression line to separate all laps into two groups: low-skilled (upper) and high-skilled (lower). We then focused on laps that scatter close to the regression line—i.e., that are ambiguous between high- and low-skilled—and labeled them as "middle". In this study, twenty percent of the total was not used in training data because they fell into this middle category.

3.3 Frequency Analysis by Wavelet Transform

We used wavelet transform to analyze the frequency characteristics of drivers' operations. Previous researchers have found that the operation frequency is quite different depending on a driver's skill [2]. For example, when passing through a curve, a high-skilled driver can drive smoothly while a low-skilled one tends to be wobbly. These characteristic differences can be hidden in the different frequency components of their operations. In order to uncover them, we added discrete wavelet transform (DWT) to the proposed technology to decompose the data on drivers' operations into different frequency components. We expected this to enable the extraction of driving operation features by the comparison of similar frequency components between laps.

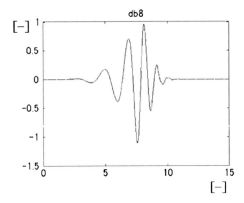

Fig. 4 Mother wavelet: db8

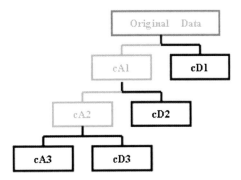

Fig. 5 Tree decomposition of wavelet transform

The first challenge with using DWT is to select the appropriate mother wavelet. Researchers have proposed many methods for such a task, most of which favor selection on the basis of the characteristics of the original signal [8]. Daubechies wavelet (db8; shown in Fig. 4) has similarities with various drivers' operation characteristics, such as a slight oscillation, and so we selected it as the mother wavelet in this study.

The tree decomposition of the wavelet transform is shown in Fig. 5. Original data is decomposed into a relative low frequency component cA1 and high frequency component cD1 in the first stage. Next, cA1 is decomposed into cA2 and cD2. We iterate the decomposition process so that one signal can be broken down into different frequency components. The output of the wavelet transform is cD1–cD3 and cA1.

The second challenge is to decide the level of decomposition and to select the most important frequency components of the driving operations. We performed an experiment to determine the high-frequency limit. A driver drove at 60 km/h on the course shown in Fig. 2b while intentionally steering with a high frequency. The steering angle sketch is shown in Fig. 6a. The FFT coefficient in Fig. 6b shows that the main frequency of the steering angle is below 5 Hz. Furthermore, D1–D8 are high-frequency components decomposed by the wavelet in Fig. 7c.

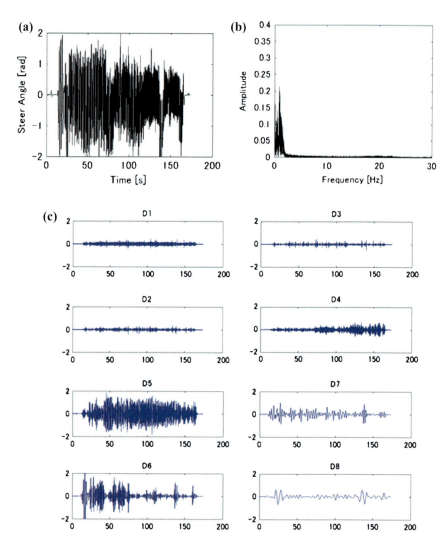

Fig. 6 (a) Steering angle. (b) FFT result of steering angle. (c) High-frequency components

The frequency becomes lower from D1 to D8. The high-frequency steering is almost completely included in D4 and D5. The frequency of D3 is 5 Hz. Consequently, after we synthesized the FFT and wavelet transform results, 5 Hz was selected as the high frequency limit.

Typically, the length of a curve and the travelling speed are used to determine the low-frequency limit. In this study, we focused on driving operation when navigating curves. Consider a curve with a length of L and a vehicle with a

Dominant Driving Operations

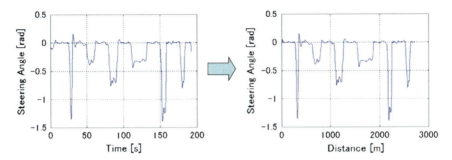

Fig. 7 Data before/after normalization

Fig. 8 Concept of window function

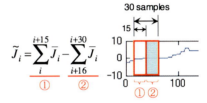

velocity of v. If the profile of an operation vibrates just one time, the cycle time is L/v. The low-frequency limit is therefore determined by

$$f = v/L \tag{2}$$

3.4 Normalization

For the purpose of comparing operation at the same location between laps, all decomposed data were normalized by distance. Because the average velocity of each lap is different, it is difficult to compare between laps using time history data. We therefore used linear interpolation to normalize all frequency components. An example of the normalized data is shown in Fig. 7.

3.5 Candidate Features

Candidate features were determined by using a sub-window function. All candidate features must reflect all the characteristics in an interval, so we added windows in the normalized data. Each window contained 30 sampling points, which corresponds to a real distance of 9 m (Fig. 8). The feature is the sum of a variable in the latter half rectangle subtracted from the sum of the former half [3].

Table 1 Training database selection results for curve 1

Driver	1	2	3	4	5	6	7	8	9	10	11	12	13	14	15	16
High-skilled	0	10	2	5	3	10	1	10	0	0	2	4	10	4	5	9
Low-Skilled	8	0	4	3	0	0	8	0	8	9	4	4	0	2	2	0

Table 2 Drivers' information. (*unit: times/month)

No.	Period of license acquired	Driving frequency	No.	License acquired period	Driving frequency
1	4 years	0	9	3.5 years	1
2	20 years	5	10	6 years	0.5
3	19 years	0.2	11	3 years	1
4	3 years	1	12	3 years	4
5	10 years	0.4	13	24 years	20
6	11 years	8	14	1 years	1
7	4 years	0.5	15	5 years	0.1
8	7 years	7	16	13 years	1

3.6 Driving Feature Extraction

A large number of candidate features were input into the AdaBoost algorithm and a few relevant features were selected. In AdaBoost, which is a machine learning algorithm widely used in face detection [4], the weak classifier is restricted to depend on only one feature and only one weak classifier is selected for each stage. Generally, the weight of each lap is the same at first. Then we train the first weak classifier, which can obtain the lowest summation of weights corresponding to laps that are incorrectly classified. In the next step, the weights for all the laps are updated and the weights for incorrectly classified laps become relatively larger and are emphasized in the next step. The second classifier is then trained and the process iterated until a combination of the extracted weak classifiers is able to separate the two groups. The features and corresponding threshold quantities contained in weak classifiers are the critical driving features between high- and low-skilled drivers.

4 Results and Discussion

4.1 Result of Training Groups Selection

The high- and low-skilled driving laps are selected by the definition proposed in the preceding section (Fig. 3). The curves in Fig. 2b are labeled 1–6 by travelling direction. Here, we consider curve 1 as an example of the results. The boundaries of curve 1 are 50 m ahead of the beginning and 50 m behind the end of the arc.

Dominant Driving Operations

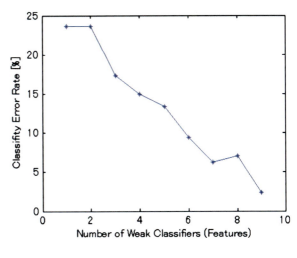

Fig. 9 Features versus error rate

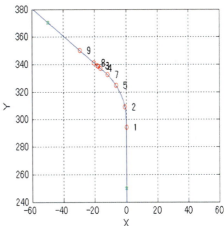

Fig. 10 Locations at which features appeared

The selection results for each driver are summarized in Table 1. Of the total 160 laps, 127 were selected for the training database: 75 high-skilled and 52 low-skilled. As shown in Table 1, most laps from drivers 2, 6, 8, 13, and 16 were determined to be high-skilled. As shown in Table 2, those drivers had possessed their licenses for a longer period than the others, and their daily driving frequency was also higher. This suggests that the proposed definition is a sensible choice.

4.2 Candidate Features

Steer angle, accelerator and brake pedal position, steering speed, and accelerator and brake operating speed were used to determine candidate features. These data

Table 3 Feature details. (* distance from the start point)

No.	Feature's name	Distance [m]*	Center frequency[Hz]	Threshold	High skilled
1	Accelerator operation speed	44.1	Below 0.156	−9.8143	Greater
2	Steer operating speed	59.1	2.5	−0.11423	Less
3	Steer operating speed	94.2	0.313	4.8063	Less
4	Accelerator pedal position	91.5	0.156	−3.9151	Less
5	Accelerator pedal position	75.6	0.156	3.8688	Less
6	Steer angle	93.6	0.313	0.3128	Greater
7	Brake pedal position	85.2	1.25	0	Greater
8	Accelerator operation speed	97.5	0.625	−0.40126	Greater
9	Accelerator pedal position	110.1	0.313	1.9317	Less

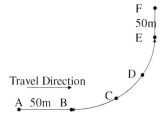

Fig. 11 Curves sections

were broken down into different frequency components and normalized by distance, after which we added a sub-window function to create candidate features.

4.3 Feature Extraction Results

Following the relationship between the weak classifier number and the error rate in Fig. 9, which is one of the most import relationships in the AdaBoost algorithm to decide the necessary number of weak classifiers, we find that the error rate reaches 3 % at nine weak classifiers. In this research, the feature selecting procedure finished when the classification rate reached 97 %. This means that nine features are necessary to get a correct 97 % classification rate. The locations at which features appeared are labeled in the order of priority on the route in Fig. 10. For curve 1, the difference between the high- and low-skilled groups was obvious at the end of the curve. The details of these features are summarized in Table 3.

4.4 Distribution of Features

To clarify the location distribution of the extracted features and the feature details listed in Table 3, we here discuss the extraction results of all six curves. First, we

Fig. 12 Curves sections

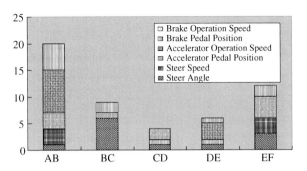

Fig. 13 Original data for accelerator pedal position (fifth feature in Table 3)

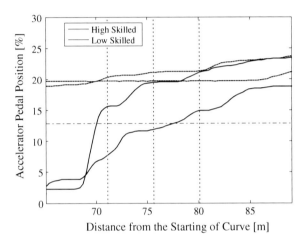

divide all curves into five sections (Fig. 11): straight sections AB and EF and sections of equal arc BC, CD, and DE.

The distribution of all extracted features from the six curves in Fig. 12 shows that accelerator and steer operations were most frequently selected. Differences between these two operations show the gap between high- and low-skilled drivers in the course. Most of the features appeared at the entrance and exit of curves, which makes sense because operational adjustments usually occur in such sections. In contrast, all drivers tended to have less operational alterations in the centre of the curves.

The fifth feature in Table 3 is the accelerator pedal position 75.6 m from the start point of curve 1. Normalized data from five laps (Fig. 13) clearly show that high-skilled drivers operate the accelerator pedal smoothly while low-skilled drivers are more hesitant: they step on the pedal, pause, and then press down. Such hesitations are particularly clear in the 0.152 Hz components.

5 Conclusion and Future Work

We proposed a statistical method technology to extract driving features between high- and low-skilled drivers from a huge database. Experiments on driving operations while passing through curves were performed using a driving simulator on a specially designed course. The proposed technology was used to extract features related to each curve.

Results showed that most features appeared at the beginning and end of curves, which makes sense as this is where operations occur most frequently. We also found that low-skilled drivers tended to operate with more hesitation.

At this stage, we have performed basic research on the extraction of driving characteristics. In the next stage, we will examine the more complex traffic environments that typically surround vehicles. We also intend to develop a driving assist system that advises low-skilled drivers via the extracted features.

Acknowledgments The author Li was supported through the Global COE Program, "Global Center of Excellence for Mechanical Systems Innovation," by the Ministry of Education, Culture, Sports, Science and Technology.

References

1. Murphey YL, Milton R, Kiliaris L (2009) Driver's style classification using jerk analysis. Computational intelligence in vehicles and vehicular systems, CIVVS '09. IEEE Workshop 2009, 23–28
2. Zhang Y, Lin WC, Steve Chin Y-K (2010) A pattern-recognition approach for driving skill characterization. IEEE Trans Intel Transp Syst 11(4):905–916
3. Li S, Yamaguchi D, Sato Y, Suda Y, Hirasawa T, Takeuchi S, Yoshioka S (2011) Differentiating skilled and unskilled drivers by using an adaboost classifier for driver's operations. In: 18th world congress on intelligent transport systems, 2011. TS92-3044
4. Viola P et al (2001) Robust real-time object detection. Int J Comput Vision 57(2):137–154
5. Mao J, Jain AK (1995) Artificial neural networks for feature extraction and multivariate data projection. IEEE Trans Neural Netw 6(2):296–317
6. Watanabe S, Furuhashi T, Obata K, Uchikawa Y (1993) A study on feature extraction using a fuzzy net for off-line signature recognition. In: Proceedings of 1993 international joint conference on neural networks. IJCNN '93-Nagoya, vol 3, pp 2857–2860
7. Lerner B, Guterman H, Aladjem M, Dinstein I (1996) Feature extraction by neural network nonlinear mapping for pattern classification. In: The 13th international conference on pattern recognition, ICPR13, 1996. vol 4, pp 320–324
8. Chinmaya K, Mohanty AR (2006) Monitoring gear vibrations through motor current signature analysis and wavelet transform. Mechanical systems and signal processing, vol 20, pp 158–187

Cross-Linking Driver Assistance Systems Via Centralized Scene Interpretation Using The Example Of Directional Roadway Detection

Markus Hörwick, Loren Schwarz, Stefan Holder, Martin Buchner and Hariolf Gentner

Abstract Modern automobiles are increasingly equipped with Advanced Driver Assistance Systems (ADAS), which improve passenger safety and comfort. Each of these ADAS can rely on a multitude of environmental sensors, which creates the need of cross-linking ADAS with respect to their access to environment information. This idea is reflected in the so-called environment model (EM) that centralizes the access of all ADAS functions to the perceived but not yet interpreted scene in the vehicle environment. In this paper, we propose an architectural framework that concretizes the EM in terms of separate modules for signal processing and sensor data fusion, as well as well-defined generic input and output interfaces. Moreover, we extend the EM by including additional components and output interfaces for centralized scene interpretation (CSI). The architectural framework allows streamlining the development process and increasing computational efficiency, as software modules for scene interpretation can be shared among multiple ADAS. In addition, the consistency of the overall ADAS behavior is improved, as all ADAS functions rely on the same scene information and scene interpretation. We illustrate our concept of centralized scene interpretation using the case study of a module for the detection of directional roadways, a generalization of motorways. The algorithm we present analyzes multiple perceptual indicators in favor or against a directional roadway and computes an overall decision, along with a foresight value for the expected duration of the directional roadway, to be used in various ADAS functions.

F2012-I01-018

M. Hörwick (✉) · L. Schwarz · M. Buchner · H. Gentner
BMW AG, Munich, Germany

S. Holder
BMW Car IT GmbH, Munich, Germany

Keywords Driver assistance systems · Environment model · System architecture · Scene interpretation · Directional roadway detection

1 Introduction and Objective

The increasing need for mobility in civil road traffic and the corresponding increasing vehicle density in the road network lead to an elevated risk of collisions caused by driver overload and inattention, as well as it causes annoyance, and psychic and physical stress. Thus, it is necessary to improve safety and comfort of modern automobiles. Advanced Driver Assistance Systems (ADAS) can make a substantial contribution to this objective.

Today, there are two main challenges concerning the development of future ADAS at BMW and throughout all major OEMs. The first one is to engineer parking and driving functions that fulfill the complete continuous longitudinal and lateral vehicle guidance. Examples are the remote controlled parking function [1], which enables an automatic parking maneuver of a driver-less car, and the congestion assistant, which enables highly automated driving in traffic jams [2, 3]. The second challenge concerns ADAS that enable automatic emergency swerve maneuvers, like lateral collision avoidance [4]. Such ADAS engage with the lateral vehicle guidance in very critical situations where the driver seems unable to avoid a collision. Overall, the vision of the ADAS community is to enable accident-free driving that is fully automated in certain situations.

Regarding the fact that modern vehicles are equipped with an increasing number of ADAS, it becomes clear that a cross-linking of all ADAS functions among each other is necessary. On the one hand, this affects the coordination and arbitration of concurrent control demands, generated by the different ADAS functions, for both the (longitudinal and lateral) vehicle guidance and the human–machine interface. As, on the other hand, the realization of each of the functions mentioned above requires more than a single physical sensor and even different sensor principles, a centralized environment perception has to be established. This paper focuses on the latter aspect.

To achieve an overall system that is highly cross-linked in the perception domain and technically controllable, and to ensure an economical development process, an efficient architectural configuration is essential. In particular, a modular design approach is required which structures the overall system in suitable software components with consistent interfaces. This way, it enables the reuse of technically mature components, as well as an efficient, economical development process by means of parallelization. In this context, it is especially important to identify cross-functional algorithms, which are executed in central modules and which provide subsequent modules with their results. As shown in Fig. 1, the cross-functional part of the perception domain is typically referred to as environment

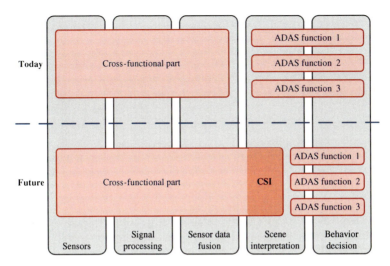

Fig. 1 Cross-functional parts of driver assistance systems today and in the future

model (EM) [5, 6]. The EM provides a data interface for all ADAS functions, which describes the non-interpreted scene in the vehicle's environment.

In this work, we make the following two contributions:

(1) We concretize the term EM in terms of an architectural framework, called EM framework, that contains modules for signal processing and sensor data fusion, as well as well-defined generic input and output interfaces. Moreover, we extend the established notion of an EM by including additional components and output interfaces for centralized scene interpretation (CSI), see Fig. 1. In doing so, the following additional main advantages become effective:

- The temporal and functional consistency of the overall ADAS behavior can be improved, as all ADAS functions do not only rely on the same scene information but also on the same scene interpretation.
- The overall required computational resources are reduced because CSI algorithms are computed centrally and results are provided to multiple ADAS functions.

(2) As an exemplary case study, we propose an algorithmic concept of a particular CSI component, the so-called directional roadway detection, which is able to recognize whether the vehicle is on a road comparable to a motorway or not.

The paper is structured as follows: first, we present the EM framework. Afterwards, the directional roadway detection algorithm is described. Finally, we draw a conclusion and give an outlook to our further research.

2 Architectural Framework

In this section, our proposed EM framework is described in detail. The goal of the EM framework is to generate information for ADAS functions and to provide access to this information via well-defined interfaces. In particular, the information about the non-interpreted vehicle environment, called scene, is generated from sensor data and provided by the scene interface. The CSI then interprets the scene information by analyzing the current situation and provides the results via the CSI interface. All ADAS functions then have access to both, the scene interface and the CSI interface. These two interfaces are designed with the goal in mind that their information can be used by as many ADAS functions as possible.

Figure 2 gives an overview of the logical layers of functional software modules, structured according to logical information flow. The logical layers are the *physical sensors layer*, the *signal processing layer*, the *sensor data fusion layer*, the *scene interpretation layer* (including function-specific and centralized scene interpretation), and the *overall situation assessment and behavior decision layer*. The separation of layers follows the state of the art [5–7]. As shown in Fig. 2, the EM framework contains sensor data fusion modules (Fig. 2 shows only one exemplary sensor data fusion module), the CSI as well as the fusion interface, the scene interface and the CSI interface. Moreover, ADAS functions are separated into function-specific scene interpretation and overall situation assessment and behavior decision. CSI and function-specific scene interpretation are logical blocks that consist of several software modules. It should be noted that the software modules of all logical layers, in particular the software modules of the environment model, are distributed over various electronic control units. The EM framework therefore defines a *distributed* environment model.

The central element of the EM framework is the scene interface (see Fig. 2), which provides a non-interpreted description of the vehicle environment. For ADAS functions, the scene interface serves as a single point of truth regarding vehicle environment information. Consequently, ADAS functions do not have access to information from layers below the scene interface. In the EM framework, the information from the scene interface is separated into six categories [8]: dynamic objects (e.g. vehicles), free space (space that is explicitly classified as being free), ground markings (e.g. lane markings), traffic signs, road surface (e.g. used for adaptive vehicle control) and predictive road information (e.g. from digital map or Car-2-X communication).

The information of the scene interface is generated as follows. The raw sensor data of a single sensor is processed by possibly multiple signal processing modules. For instance, the raw images of a camera may be processed by a lane detection and an object detection module. If the information of a scene category is provided by a single signal processing module (category *a* in Fig. 2, e.g. lane markings are obtained only from the camera lane detection module), then the result of the signal processing module is directly made accessible by the scene interface. In contrast, if the information of a scene category is provided by multiple signal

Cross-Linking Driver Assistance Systems

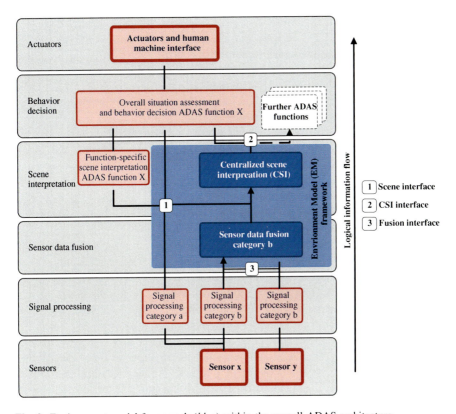

Fig. 2 Environment model framework (*blue*) within the overall ADAS architecture

processing modules (category *b* in Fig. 2, e.g. dynamic objects are provided by the object detection of the camera and the radar), then the redundant information is fused to obtain a single and consistent representation of each scene category. Since sensor data fusion modules are essential to provide consistent environment information, these modules, along with their input interfaces (fusion interface in Fig. 2), also belong to the EM framework.

In the scene interpretation layer, the non-interpreted information from different scene categories is combined and interpreted. The individual scene interpretation modules are separated into function-specific scene interpretation and centralized scene interpretation (CSI). Each function-specific scene interpretation module generates information for a single ADAS function only and is therefore part of the corresponding ADAS function. In contrast, the CSI modules generate information for multiple ADAS functions. The CSI and its output interface (CSI interface in Fig. 2) are therefore an essential part of the overall EM framework. It is important to note that ADAS functions still have access to the entire scene interface, in addition to the CSI interface.

The logical block CSI can be further separated into the following logical sub layers: fusion of physical sensor information and predictive road information, situation analysis, and situation assessment. In the first sub layer, redundant information from physical sensors and predictive road information is fused to obtain a single, consistent scene representation. This increases the quality and availability of the fused information. The fusion of physical sensor information and predictive road information is part of the CSI since the sensor data fusion layer only fuses data from physical sensors. An example module for this CSI sub layer is the fusion of road geometry from digital map and camera-based lane detection, called curve foresight [9, 10]. The goal of the situation analysis sub layer is to interpret the so far non-interpreted environment information from the scene interface to identify all situation aspects that are relevant for subsequent situation assessment and behavior decision. One example for a situation analysis CSI module is the construction site detection, which detects whether or not the ego vehicle is approaching a construction site. Another example is the directional roadway detection, which identifies whether or not the ego vehicle is on a roadway that is physically separated from the oncoming roadway, as in the case of a motorway (see below). Finally, the situation assessment sub layer has the goal of identifying decision alternatives and objects that are relevant for the subsequent behavior decision. An example for a situation assessment module is the logical driving corridor, which is defined by lane markings, road boundaries, and other vehicles on the ego lane and its neighboring lanes [5].

The overall situation assessment and behavior decision layer is the core part of each ADAS function. This layer consumes the information provided, on the one hand, by the EM framework via the scene interface and the CSI interface, and on the other hand, by the function-specific scene interpretation. Based on the overall situation assessment, the appropriate behavior is determined and responses of the actuators and human machine interface are triggered.

3 Directional Roadway Detection

In this section, we describe our proposed algorithm for directional roadway detection, as an example for a situation analysis CSI module. This algorithm has the purpose of recognizing whether the vehicle is on a directional roadway or not, and in addition, to estimate the distance until a directional roadway ends or begins, respectively. We define a directional roadway (DRW) to be a road where all traffic within constructional, un-passable boundaries is moving in the same direction. A motorway is a special case of a directional roadway. The crucial aspect of a directional roadway is that no oncoming traffic needs to be expected on the neighboring lanes. Although digital map data can provide information on the current road type, this information is not always available and can be corrupted by measurement errors. In addition, roadway changes might not be reflected in out-dated maps, such that a validation of the roadway information based on evidence from the scene interface, as proposed here, can be highly beneficial.

Table 1 ADAS functions using the information provided by the DRW detection module

ADAS function	Purpose	Need for DRW detection
Swerve assistant [11]	To prevent frontal impacts by braking and steering the vehicle sideways onto another lane	Knowledge about the directional roadway status (and the expected distance to a change) is vital in order to prevent hitting oncoming traffic on a neighboring lane
Lateral collision avoidance [4]	To prevent lateral impacts by steering the vehicle sideways onto another lane, if necessary	
Congestion assistant [2, 3]	To control the vehicle fully autonomously up to a maximum speed in traffic jam situations	Functionality is restricted when not on a motorway and thus reliable directional roadway detection (and information on the expected distance to a status change) is required
Lateral and longitudinal adaptive cruise control	To control the vehicle by keeping a user-defined speed and enforcing safe distances to other vehicles, including longitudinal and lateral control	

The key idea of the algorithm is to combine the contribution of multiple perceptual input indicators into one aggregated directional roadway decision. Each of the input indicators alone is too weak for a decision, but together the indicators validate each other. The input indicators are computed from signals provided by various sensing modalities, including camera data, GPS tracking with a digital map and vehicle dynamics. In accordance with the centralized scene interpretation architecture, the directional roadway detection module provides its output to multiple different ADAS. Referring to Fig. 2, the DRW module belongs to the CSI block of the scene interpretation layer (sub-layer situation analysis).

3.1 Requirements and System Architecture

We collected requirements from the function owners that are responsible for the development of individual driver assistance systems. Each of these functions has a particular reason for requiring directional roadway detection information, as summarized in Table 1.

The following non-interpreted signals, originating from sensory systems, are used as inputs: stereo front cameras (e.g. object tracking, free space detection, lane marking detection, and traffic sign recognition), digital map (e.g. road type, lane count and marking information, road curvature, segment length) and vehicle dynamics (e.g. speed, acceleration, steering angle). The outputs of the DRW detection module are a Boolean directional roadway flag d, a confidence value c and an extrapolated foresight value s that gives an estimate of how long the current directional roadway state will be valid, assuming that the perceptual indicators remain unchanged.

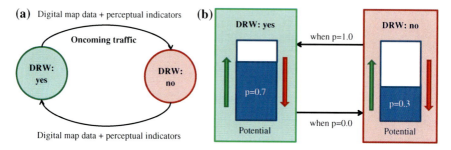

Fig. 3 a State diagram with main triggers for the directional roadway decision. **b** Internal potential computed from the perceptual indicators

3.2 Algorithm Description

The motivation for using the algorithm presented here, as opposed to common classification algorithms, is the functional requirement to provide an estimate on the distance to the beginning or end of a directional roadway in addition to the current DRW status. From a high-level perspective, our algorithm combines two preliminary internal directional roadway decisions to a final output consisting of two states (Fig. 3a): one of these internal decisions is based on the street type information given by the digital map, and the other internal decision is the result of weighting and aggregating a variety of indicators, including perceived environmental information and digital map data. Apart from the internal decisions, the detection of oncoming traffic within constructional road boundaries leads to an immediate switch to the no DRW state.

Let $d_{MAP} \in \{+1, -1\}$ denote the directional roadway decision based on digital map data that is true (i.e. +1), if the current street type matches a series of pre-specified types. Similarly, let $d_{IND} \in \{+1, -1\}$ denote the decision based on the perceptual indicators. The confidence value $c_{MAP} \in [0, 1]$ is extracted from the digital map system and indicates how well the current GPS position can be matched to a map position. The confidence value $c_{IND} \in [0, 1]$ is based on the aggregation of the perceptual indicators, as described below. We define the final decision of the DRW detection module to be $d = sign(r)$, with an overall confidence $c = |r|$. Here, the value $r \in [-1, +1]$ is the result of averaging (with equal weights) the decision and confidence of digital map and perceptual indicators:

$$r = \frac{1}{2} c_{MAP} \cdot d_{MAP} + \frac{1}{2} c_{IND} \cdot d_{IND} \qquad (1)$$

Each perceptual indicator represents one criterion that can positively or negatively influence the directional roadway decision. Every indicator is based on one or multiple un-interpreted input signals from sensory systems that are combined in simple fusion algorithms. Examples of perceptual indicators we use include the following:

Cross-Linking Driver Assistance Systems

- Number of directed lanes (digital map and camera data).
- Width of lanes (camera data).
- Road curvature (digital map and camera data).
- Map segment length (digital map data)/.
- Vehicles driving in parallel (camera data).
- Smoothness of road boundaries (camera data).

Aggregation of the perceptual indicators is based on two main concepts: the potential and the verification distances. The potential $p \in [0, 1]$ is an accumulator that is updated in every time step Δt according to $p = p_{old} + \Delta p$, where p_{old} is the potential in the previous time step and Δp is an incremental update. As shown in Fig. 3b, the preliminary internal directional roadway decision d_{IND} based on the perceptual indicators is set to $+1$ as soon as $p = 1$, and to -1 as soon as $p = 0$.

For each perceptual indicator identified with an index i, we define a function f_i that maps a given value $x^{(i)}$ of this indicator to a real value $f_i(x^{(i)})$ that we refer to as its verification distance. Intuitively, the verification distance is the distance the vehicle needs to travel such that the potential is either filled from 0 to 1, or emptied from 1 to 0, given the current value of the indicator and assuming this indicator is focused alone. The potential update is computed as a sum of contributions from all perceptual indicators:

$$\Delta p = \sum_i \Delta p_i = \sum_i \frac{1}{f_i(x^{(i)})} \cdot v \Delta t = \frac{v \cdot \Delta t}{f_{ALL}}. \qquad (2)$$

Here, v is the vehicle speed during the last time step and the numerator is thus the distance travelled. In the summation, each perceptual indicator contributes in an inverse proportion to its verification distance. Thus, a large verification distance for a given value of an indicator results in a small contribution of the indicator during one time step. The potential is saturated at the value of 1 (and 0), so that any positive (or negative) updates Δp_i are disregarded in these cases, respectively.

Verification distances have to be pre-specified statically for each perceptual indicator. Each potential value of a perceptual indicator needs to be assigned a suitable verification distance. In the case of indicators with a continuous value range, verification distances can be specified for several indicator values and interpolated in-between (see Fig. 4). Verification distances for several indicators can be extracted systematically using a statistical evaluation of map data. Other indicators require heuristics to determine suitable verification distances.

We propose to use the level of the potential p to provide a measure of certainty for the internal directional roadway decision of the perceptual indicators. This internal confidence value is denoted as c_{IND} and is used in Eq. 1, where the decision of the digital map system is combined with that of the perceptual indicators. We define the confidence as

$$c_{IND} = \begin{cases} p & \text{if } d_{IND} = +1, \\ 1 - p & \text{if } d_{IND} = -1. \end{cases} \qquad (3)$$

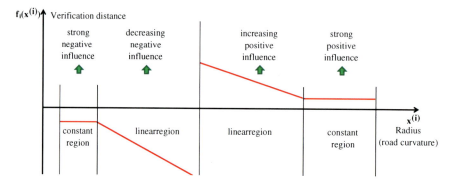

Fig. 4 Verification distances for an exemplary perceptual indicator (road curvature)

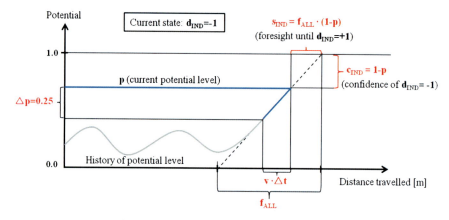

Fig. 5 Exemplary potential with current update and resulting confidence and foresight

The verification distances of all perceptual indicators and the resulting potential level also allow us to compute an internal foresight value. This value, denoted as s_{IND}, represents the expected distance to be travelled until the directional roadway ends (if $d_{IND} = +1$), or until a directional roadway begins (if $d_{IND} = -1$). We define the extrapolated foresight value as

$$s_{IND} = \begin{cases} f_{ALL} \cdot p & \text{if } d_{IND} = +1 \text{ and } \Delta p < 0, \\ f_{ALL} \cdot (1-p) & \text{if } d_{IND} = -1 \text{ and } \Delta p > 0, \\ \infty & \text{otherwise.} \end{cases} \quad (4)$$

The value f_{ALL} is introduced in Eq. 2 and represents the verification distance combined over all individual perceptual indicators, i.e. the estimated distance to be travelled to fill or empty the potential entirely, from 0 to 1 or from 1 to 0, respectively.

An example for a situation where the foresight is set to infinity is when $d_{IND} = +1$ and $\Delta p > 0$. In this case, the potential evolves in a direction that supports the decision $d_{IND} = +1$ and an estimate of the end of the DRW is thus not possible. Figure 5 illustrates how the confidence c_{IND} and foresight s_{IND} are computed from the potential level, assuming the vehicle is currently not on a directional roadway.

Similar to the confidence value, we also combine the internal foresight of the perceptual indicators with the foresight based on the GPS and map information. Let s_{MAP} be the foresight based on the digital map system, given by the distance along the most probable path until the road type changes. Depending on the final directional roadway decision d, the final foresight value is $s = \text{Min}(s_{MAP}, s_{IND})$, if $d = +1$, and $s = \text{Max}(s_{MAP}, s_{IND})$, if $d = -1$.

4 Conclusion and Outlook

This work concretizes the notion of the term environment model (EM) by proposing an architectural framework, called EM framework. The framework defines modules for signal processing and sensor data fusion, as well as centralized scene interpretation, which work together via well-defined generic interfaces to provide consistent information about the vehicle environment to all ADAS functions in a car. The EM framework serves as a single point of truth concerning environmental information for current highly cross-linked ADAS and thereby enables an overall system that is technically controllable and that facilitates an economical development process. The integration of scene interpretation modules into the EM framework is a novel approach that makes the environmental perception even more consistent and effective. To make this aspect more tangible, we presented the algorithmic concept of the directional roadway detection as an exemplary scene interpretation module. This module recognizes whether the vehicle is on a road comparable to a motorway.

We believe that standardizing the EM framework will be highly beneficial for the entire automotive industry. In particular, well-defined interfaces between environment perception, scene interpretation and ADAS functions, as proposed in this work, would facilitate the distributed development process across OEMs and suppliers.

In the future, first the DRW algorithm has to be extensively evaluated. In addition, we will investigate whether machine learning methods, e.g. support vector machines, are applicable to improve the performance of the rule-based approach proposed in this paper. In general, we will also identify and implement additional CSI modules. Several first steps have already been taken, as we are currently starting to work on cross-functional modules for logical driving corridor, curve foresight and construction site detection.

References

1. BMW AG Active PDC and remote controlled parking. http://www.connecteddrive.info/index.php?option=com_content&view=article&id=76, Accessed on 31/05/2012
2. Schaller T (2009) Congestion assistance—supporting the driver in lateral and longitudinal guidance in congestion situations. Dissertation at Technical University of Munich, Garching, 2009
3. BMW AG Traffic jam and queuing assistant. http://www.connecteddrive.info/index.php?option=com_content&view=article&id=73. Accessed on 31/05/2012
4. BMW AG Lateral collision avoidance. http://www.connecteddrive.info/index.php?option=com_content&view = article&id = 70. Accessed on 05/31/2012
5. Hörwick M (2011) Safety concept for highly automated driver assistance systems. Dissertation at Technical University of Munich, Garching, 2011
6. Holder S, Hörwick M, Gentner H (2012) Funktionsübergreifende Szeneninterpretation zur Vernetzung von Fahrerassistenzsystemen. Braunschweiger Symposium Automatisierungssysteme, Assistenzsysteme und eingebettete Systeme für Transportmittel (AAET)
7. Maurer M (2000) Flexible Automatisierung von Straßenfahrzeugen mit Rechnerschen. Dissertation at University of Bundeswehr München, Neubiberg, 2000
8. Schöttle M (2011) Zukunft der Fahrerassistenz mit neuen E/E-Architekturen. ATZ Elektronik, 4/2011
9. Tsogas M, Polychronopoulos A, Amditis A (2007) Using digital maps to enhance lane keeping support systems. In: IEEE intelligent vehicles symposium (IV), Istanbul, 2007
10. Gackstatter C, Heinemann P, Thomas S, Rosenhahn B, Klinker G (2010) Fusion of clothoid segments for a more accurate and updated prediction of the road geometry. In: IEEE conference on intelligent transportation systems (ITSC), Madeira, 2010
11. Stählin U, Schorn M, Isermann R (2006) Notausweichen für ein Fahrerassistenzsystem zur Unfallvermeidung. Wiesloch, Fachtagung Steuerung und Regelung von Kraftfahrzeugen und Verbrennungsmotoren (AUTOREG)

Part II
V2X Communication Technology

Towards Design and Integration of a Vehicle-to-X Based Adaptive Cruise Control

Oliver Sander, Christoph Roth, Benjamin Glas and Jürgen Becker

Abstract Today, assistance systems more and more rely on external environmental information. This information exchange is widely still limited to on-board sensors and the local area around the vehicle. Considering e.g. typical safety systems, information is gathered via local sensors. If a critical state is detected, existing actuators are respectively triggered. Vehicle-to-Vehicle or Vehicle-to-Infrastructure (Vehicle-to-X, V2X) Communication allows breaking these limitations, since vehicles are able to exchange internal information with other vehicles wirelessly. The single vehicle is then able to communicate with its environment beyond the line of sight. This results in an earlier detection of critical traffic situations, increased traffic safety and an optimized traffic flow. While currently major efforts regarding V2XC are taken to standardize protocols and communication, integration of V2XC into the overall electric/electronic (E/E) architecture of a vehicle, let alone integration aspects of combining V2X technology with established vehicular safety systems has not been looked at in detail. Using V2XC for safety critical applications demands for strict adherence to real time constraints in the overall processing chain, which starts at the sensor in the transmitting vehicle and ends up in actuators of the receiving vehicles. In between a lot of necessary processing steps make meeting latency limits a major challenge. Looking especially at security checking, data processing and data aggregation we implemented an FPGA-based approach for a V2XC system that is able to fulfil the upcoming computational demands. The system is tightly coupled to the automotive E/E architecture by extending a central car

F2012-I02-002

O. Sander (✉) · C. Roth · J. Becker
Institute for Information Processing Technology (ITIV), Karlsruhe Institute of Technology (KIT), Karlsruhe, Germany

B. Glas
Robert Bosch GmbH Embedded Security CR/AEA3, Stuttgart, Germany

gateway. Within this work we present our approach of combining V2X technology with Adaptive Cruise Control (ACC) while utilizing the aforementioned V2X system as starting point. The paper is meant to provide a basis for future realization of V2X-based safety systems and their tight integration into E/E architectures. Therefore, we explain most important aspects concerning combination of our centralized V2X approach together with ACC capabilities within a real prototype car. This includes a declaration of the underlying concepts as well as a discussion of design decisions that were made to interface ACC and V2X with respect to performance, security and safety requirements.

Keywords Vehicle-to-vehicle/vehicle-to-x communication · Adaptive cruise control · Electronic architectures · Prototyping

1 Introduction

One of the basic innovation drivers in the automotive industry has been the improvement of safety. A next major contribution to this goal is the usage of direct communication between vehicles and also to enhanced infrastructure—collectively denoted as V2X (vehicle-to-X) communication. Exchange of real-time information about traffic and road conditions, but also of basic information about position and movement of traffic participants provides a more detailed view of the vehicles surroundings beyond direct line-of-sight. This enables more precise and faster driver assistance on one hand, and on the other substantially improved traffic efficiency due to more exact, up-to-date information about traffic density and flow.

Extensive research activities and numerous national and international projects and consortia for investigation, road trial (cf. e.g. simTD [1]), harmonization (including e.g. COMeSafety [2] and the car-to-car communication consortium (C2CCC) [3]), and standardization (for example by IEEE [4] and SAE [5]) brought the field to a level of knowledge and coordination that makes introduction of first systems and functions in the current decade not only possible, but probable.

Work in the area of possible applications and protocols and concerning architecture of the overall system are substantial and advanced, but especially in the area of realization on available hardware and integration in the local vehicle system remain interesting open issues, questions and challenges. For example real-time capability of the system considering communication and processing is fundamental for its performance in critical situations. At the same time the new communication ability between vehicles has to be integrated in the vehicle-internal communication infrastructure.

In the following a system approach based on reconfigurable hardware, especially field-programmable gate arrays (FPGAs), is presented that offers solutions for realization and integration. After a short presentation of the overall architecture and its main properties, an approach for the realization of a V2X based Adaptive Cruise Control system is presented in detail.

2 Motivation and Challenges

2.1 Adaptive Cruise Control State of the Art

An Adaptive Cruise Control (ACC) is an automotive speed control system which considers the distance to the vehicle in front as an additional control variable for automatic distance adjustment [6]. ACC systems are mainly used with automatic transmission since transmission is done by the gearbox autonomously. The main goal is an increased comfort and a relief of the driver's concentration during long freeway journeys. A second main goal is an increased safety e.g. by provision of acoustic or optical warnings in case of low distances or automatic preparation and execution of emergency braking. Generally, an ACC determines position and speed of the vehicle in front with a special sensor. Based on the received data, speed and distance of the following vehicle are adaptively controlled using motor and brake intervention. In today's vehicles mainly radar sensors are applied. Radar systems typically work at 10 mW which does not constitute a risk for health. The radar frequency typically ranges between 76 and 77 GHz. Newer systems use 24 GHz which makes them cheaper. Beside radar based systems, there also exist lidar (light detection and ranging) systems. However, they currently exhibit too high disturbances in case of a limited sight distance due to bad weather conditions. The advantage of lidar systems is their more favourable price. They typically work in the infrared band which is invisible for the human eye.

The most obvious limitation of all currently available ACC is their sensing limitation to the line of sight. This can be a safety critical factor in situations like e.g. traffic jams that can effect a delayed braking behaviour if several vehicles approach a hidden queue one after another. In that case, the systems response time is only dictated by position and speed of the vehicle in front.

2.2 V2X Research Directions

V2X communication is currently being standardized in various organizations. In the USA the IEEE [4] and SAE [5] developed draft standards. In Europe results of several groups (e.g. the Car2Car Communication Consortium (C2CCC [3]), SeVeCom [7], and ETSI [8]) are collected within the COMeSafety project [2]. Looking at the research landscape, several projects have been conducted and are still running. E.g. within the research project EVITA [9] security aspects and their integration with V2X technology has been explored. The SeVeCom project [10] examined requirements primarily from a security point of view. In the simTD project [1] the focus is on the development of a holistic V2X infrastructure for

Fig. 1 Highway layout of OVC

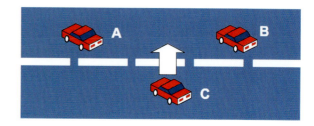

demonstration purposes and measurements by means of real scenarios. The integration of V2X functionality into the E/E architecture of a vehicle and the consideration of the complete processing chain starting at the sensor in the transmitting vehicle and ending in actuators of the receiving vehicles, is neglected within the mentioned projects. Especially using V2XC for safety critical applications demands for strict adherence to real time constraints in the overall processing chain.

2.3 V2X-ACC Example Scenario and Challenges

In addition to the traffic jam example that has shortly been described above, another example scenario called One-Vehicle Cooperative ACC (OVC) in which the usefulness of a V2V based ACC system is demonstrated is described in [11]. As illustrated in Fig. 1 vehicle B is following its preceding vehicle A. Then vehicle C in the adjacent lane cuts in between the two vehicles and becomes the new preceding vehicle.

Without V2V communication the ACC detects the cut-in vehicle when the latter passes the lane border. The ACC controller of vehicle B effects a braking which often occurs abruptly to make space in front for the cut-in vehicle. Using V2V communication vehicle C transmits a message to vehicle B at the instant it starts to cut in from the center of the adjacent lane. Vehicle B then has approximately half of the lane change time to slow down and make space for the cut-in vehicle [11]. The authors show beneficial effects of communication in both efficiency (average velocity) and cost (braking effort).

In order to realize such an application using V2X/V2V technology, a large set of performance and security constraints need to be fulfilled. E.g. the communication and processing chain from vehicle C to vehicle B roughly includes

1. latency for gathering of position data from sensors in vehicle C (given by GPS system)
2. processing latency of the OVC application in vehicle C
3. internal communication latency via bus systems in vehicle C
4. external communication latency via V2V wireless channel (including security)
5. processing latency of the OVC application in vehicle B
6. internal communication latency via bus systems in vehicle B
7. latency for driving motor and braking actors in vehicle B.

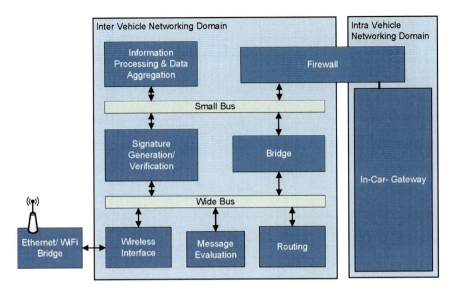

Fig. 2 V2X-gateway architecture

In order to guarantee the overall application to work correctly and to automatically provide a safe distance between vehicle C and B the whole processing chain must be below a given latency boundary. Obviously, the kind of integration of the V2V/V2X system into the E/E architecture as well as the distribution of the ACC application onto the given computation and communication infrastructure within a vehicle has great influence on system latency. Finding a good solution is a great challenge due to a large design space. Concerning the first point of E/E architecture integration of the V2X functionality, we first present our V2X-Gateway approach in the now following section. Based on this, design decisions for ACC application integration based on this E/E architecture extension are detailed afterwards.

3 V2X-Gateway Architecture

3.1 Overview

The V2X gateway architecture was previously presented in [12–14]. It is based on the last North American and European standardization efforts [2, 4, 5]. The architecture is illustrated in Fig. 2. It provides a flexible and modular FPGA-based hardware framework [15]. A central vehicle gateway has been chosen as integration point of the V2X system into the automotive E/E architecture. Thus, the system directly provides a link between the V2X network and the internal vehicular bus network. It is therefore divided into an Inter-Vehicle-Networking-Domain (InterVND and an Intra-Vehicle-Networking-Domain (IntraVND).

Typically occurring tasks are distributed on specific hardware modules (e.g. for security and data preprocessing, routing or evaluation and filtering of single incoming messages). These work in parallel and allow processing of V2X messages in discrete steps. Messages are directly passed from one module to the successor in the processing chain by means of an on-chip packet transmission bus system [13]. Each of the modules consists of a complex structure of different hardware and software components thus forming a heterogeneous multicore device. Since environmental conditions are often unforeseeable, self-adaptability significantly contributes to respond to the different demands of urban and rural traffic. Because of that self-X characteristics have been embedded into the system wherever possible e.g. by utilizing partial dynamic reconfiguration. Also trustworthiness of messages is indispensable since safety relevant decisions have to be based on remotely collected information to overcome line-of-sight limitations. Therefore, cryptographic functions have been implemented in hardware which enables efficient signature verification and secure message transmission.

3.2 Message Forwarding

Message transmission on the lower communication layers (MAC and PHY) is realized in a separate WIFI Interface module. Transmission packets are directly forwarded to the WiFi module and then transmitted over the wireless channel. Received WiFi packets are rated with a priority and immediately broadcasted to the relevant modules connected to the Wide Bus (see Fig. 2). The routing module processes the message and decides autonomously about forwarding based on appropriate routing algorithms. In order to validate the messages content the signature is verified in the signature/security module in parallel. As soon as the signature is verified and no longer needed it is discarded. If signature verification fails the message is deleted. The message is then forwarded to the information processing module (IPM). The IPM collects and aggregates information received through the VANET and prepares outgoing packets using information from on-board sensors. Communication to the intra-vehicle domain is done through a hardware firewall separating the mostly event-driven VANET side and the rather time-triggered internal bus systems. To minimize busload on these safety-critical internal structures, the IPM offers data processing up to complete applications and is therefore able to provide high-level aggregated information.

3.3 Information Processing

The IPM has to fulfil two major goals: First the execution of algorithms has to be done as fast as possible in order to meet the requirements concerning latency and throughput. Second high flexibility in system design and algorithm implementation

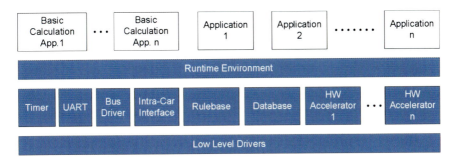

Fig. 3 Information processing module (software stack)

is of great importance. There should be the opportunity to make changes and optimizations of algorithms also in later stages of development.

The central processing element of the IPM hardware is a softcore processor on which application specific algorithms can be implemented and executed in software. Repeatedly occurring similar calculations are executed with support of special optimized hardware units which are directly connected to the processor either by a bus or as coprocessor. Looking at the software, the IPM can be structured as shown in Fig. 3. The hardware details and drivers are hidden by a hardware abstraction layer and a runtime environment (RTE) on which an arbitrary number of applications can be executed. The runtime environment provides a special standardized interface which can be accessed by the applications. The main task of the interface is to enable new applications to register themselves to the system. Registered applications are called when the database is updated with relevant information. That way a kind of decoupling of the two layers is achieved. For example future applications can easily be added to the system subsequently. The decoupling mechanism facilitates the extensibility and changeability of the software. The RTE approach can be seen similar to an RTE as proposed by the AUTOSAR consortium [16].

4 V2X-Based Adaptive Cruise Control (V2X-ACC)

4.1 V2X Based ACC

Integrating a V2X based ACC into an E/E architecture by nature gives some freedom regarding the functions location. This inherently includes, that the function maybe split into different sub-functionalities, which might be realized on one single ECU (centralized approach) or on various ECUs (distributed approach). This also depends on the fact, whether the system uses solely V2X data for the ACC, or radar data from the original sensor is also available. The radar delivers

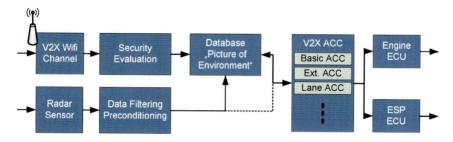

Fig. 4 V2X-ACC communication architecture

additional data for better perception of the environment especially if not all vehicles are equipped with V2X communication systems. If both data exist the V2X-ACC has to perform the data fusion on the sensor data in order to generate a consistent environmental picture. The communication of the blocks forming the V2X-ACC is depicted in Fig. 4.

The major data source in our application scenario, are the cooperative awareness messages (CAM) [2] coming from the surrounding vehicles. Each of these messages contains fundamental information about vehicle type, position and movement. As mentioned before the first step is to ensure trustworthiness of the message by means of a digital signature. If the message is trustworthy the information is copied into the picture of the vehicles environment (database).

The V2X-ACC then gathers all necessary information from the database. Obviously this includes the position, speed and acceleration of the next vehicle in front. But also the information from other vehicles that are more ahead can be used to improve the system. This allows reacting earlier on breaking vehicles that are further ahead or just around the corner. Especially in latter case such a system significantly increases safety as it uses information which is neither available for the radar nor the driver. But also vehicles that are about to perform a lane change can be taken into account very early, hence providing higher comfort and safety by earlier reactions.

If radar data is measured, this can be also stored into the aforementioned database thus complementing the environmental picture. Regarding the data fusion various realisation can be perceived. For example this could be a full integration of the data into the environmental picture. This way the V2X-ACC could not distinguish between both data sources. On the other hand radar data could be transmitted to the V2X-ACC directly. Most obvious advantage of the first solution is the enrichment of the environmental picture, usable for diverse applications.

Based on the data describing the surrounding traffic the V2X-ACC finally calculates its output data. In principal this is a request for higher or lower speed. If a higher speed is needed, the command is given to the engine management. If less speed is needed a requested to the breaking management (typically ESP ECU) is given in addition.

4.2 Design Space

The mapping of V2X system functions can be done centralized or distributed. A distributed solution might consist of a V2X communication point, which is responsible for V2X data reception and transmission including security measures. The environmental picture then is built on another ECU (e.g. a Data Aggregation ECU—DAE). Eventually the V2X ACC runs on a third ECU, getting data from the DAE. First of all such an approach would lead to massive communication between the V2X-communication unit and the DAE also introducing an additional delay. In a second step some communication link between DAE and the V2X-ACC ECU has to be established introducing additional delay.

Following the current trend of ECU consolidation one could integrate all three parts V2X communication, database handling, and V2X-ACC on one single ECU. Especially this brings the benefit of reduced communication delays between the different parts of the system. Communications to other ECUs includes driving commands to ESP and Engine Controller as well as data reception from the radar system.

A third approach could realize the environmental picture database and the V2X data handling on one ECU and the V2X ACC on the radar system (original ACC ECU). Probably this would lead to more communication compared to the aforementioned scenario. This is true because the database comprises more ACC relevant data points compared to the line of sight data from the radar system. However this implementation results in a better functional clustering.

5 Prototypical Realization and Verification

5.1 Exemplary E/E-Architecture Integration into Mercedes-Benz SL

The original radar based ACC works as follows: As soon as activated by the driver (information via CAN) the system tries to drive with the speed that has been selected by the driver. This is achieved by commands, being sent (via CAN bus) to the engine management ECU and the ESP ECU respectively. In our implementation of the V2X-ACC we are constrained by the fact of not having access to the ECU software stacks of the original vehicle. Especially this includes the original ACC ECU, the ESP ECU, the engine control unit and the format of CAN communication messages.

In order to integrate the V2X-ACC the system some modifications to the E/E-architecture have been made. The V2X-ECU follows the centralized integration scenario. It is attached to the V2X channel and three CAN channels. The first one interfaces the Interior CAN and the second one the powertrain CAN. The third CAN connection is used to connect the original ACC-ECU. This way the

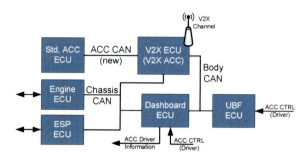

Fig. 5 V2X-ACC E/E-integration

V2X-ECU is "man-in-the-middle" regarding the ACC functionality and hence can modify all ACC data packets accordingly. This way we are able to investigate both V2X-ACC only and V2X-radar-ACC. The resulting architecture is depicted in Fig. 5.

Drivers interaction with the ACC remains unchanged. This includes, activation, deactivation, setting speed, distance etc. Information to the driver is given via display on the dashboard that shows the own vehicle and the next vehicle in the front.

5.2 ACC Application Structure

The implementation of the V2X-ACC software stack is done within the suggested software stack of the V2X system. In a first step only cooperative awareness messages (CAM) are used for the ACC system. This way the ACC system is not built upon special messages but all necessary information is derived from the standard alive beacons of the surrounding vehicles. As soon as the data packets are received in the IP-module they are sorted into the database. In the prototypical implementation a sorting mechanism is used, which in a first step distinguished lanes, in a second step rough position (front vs. back) and finally the relative distance to the receiving vehicle. This way fast and direct access to the messages of the relevant vehicles nearby is possible. For each vehicle the type, position, speed and acceleration are stored. In the current implementation only data of 100 surrounding vehicles are stored. This number has been arbitrarily chosen and can be modified.

The database is then accessed by the V2X-ACC. The ACC application is clustered into four different sub-functions, implementing a modular approach. The first function only incorporates the data of the next vehicle in the front. The second module takes into account data of vehicles, which are before this vehicle. The third one checks neighbouring lanes for entering vehicles. The fourth module assists own lane changes and looks for vehicles from behind. By using a modular approach, we are able to compare the influence of different ACC sub-functions on the overall performance of the system. Within the current implementation, we do not use the original ACC.

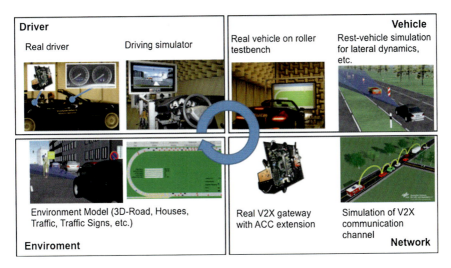

Fig. 6 V2X-ACC-in-the-loop verification environment

5.3 Model-Based Verification

In order to functionally verify the developed application we draw back to the Car2X-in-the-Loop platform that has been proposed in [17]. The base frame of the platform is given by the so called X-in-the-Loop framework [18] which describes a consistent and integrated development environment for drive systems. Thereby, "X" stands for the unit under test (UUT) which is in our case the complete vehicle equipped with V2X-ACC functionality. The Car2X-in-the-Loop platform is made up of the four domains vehicle, driver, environment and Network. Generally, each of them can be either virtual or real which allows evaluating and investigating interactions and interaction chains between real as well as virtual and real systems and sub-systems.

In Fig. 6 an instance of the platform is exemplarily illustrated that is used for evaluating the impact of the V2X-ACC on the braking behaviour of a real driver. Thereby, the driver is sitting in the real V2X-ACC vehicle which is running on a roller testbench or controls it via a driving simulator. Both, real driver and real vehicle are embedded into a virtual environment that simulates surrounding traffic on microscopic level. Furthermore, the real wireless network channel of the vehicle is connected to a virtual wireless network communication channel that allows V2XC between real and virtual vehicles.

First experimental results substantiate correct functionality of the overall V2X-ACC implementation. Performance results e.g. concerning brake behaviour and latencies are still under investigation.

6 Conclusion

In this paper an approach for combining V2X technology with ACC functionality has been presented. Based on the discussion of general realization alternatives of the V2X-ACC application, a possible realization has been derived that is tailored for our existing V2X system and that considers performance, security and safety requirements. The approach sets on top of an already existing V2X system concept that extends the central gateway control unit and therewith E/E architecture of a vehicle with V2X functionality. The chosen design decisions go hand in hand with the trend towards more centralization that can currently be observed in the automotive industry. The system has been functionally verified by a model based approach called Car2X-in-the-Loop. We hope, this work can provide a basis for future realization of V2X-based safety systems and their tight integration into E/E architectures in general.

References

1. simTD (2008) Sichere Intelligente Mobilität: Testfeld Deutschland. Project webpage. Available: http://www.simtd.de
2. COMeSafety Project (2008) European Communication Architecture FRAME Annex 10-1. Available: www.comesafety.org
3. Manifesto—Overview of the C2C-CC System v1.1 (2007) CAR 2 CAR Communication Consortium, 28 Aug 2007
4. IEEE (2006) Trial-use standard for wireless access in vehicular environments (WAVE), IEEE standard 1609
5. SAE (2006) Dedicated short range communications (DSRC) standard draft, SAE standard J2735
6. Dietsche K-H, Jäger T, Robert Bosch GmbH (2003) Kraftfahrtechnisches Taschenbuch. 25. Auflage, Friedr. Vieweg & Sohn Verlag, Wiesbaden
7. Papadimitratos P, Buttyan L et al (2008) Secure vehicular communication systems: design and architecture. IEEE Commun Mag 46(11):100–109
8. ETSI (2009) European Telecommunications Standards Institute. Available: http://www.etsi.org
9. EVITA project (2008) E-safety vehicle intrusion protected applications. Project webpage. Available: http://evita-project.org
10. SeVeCom project, Secure Vehicle Communication. Project Webpage. Available: http://www.sevecom.org
11. Qing X, Sengupta R (2003) Simulation, analysis, and comparison of ACC and CACC in highway merging control. In: Proceedings of IEEE intelligent vehicles symposium, pp 237–242, 9–11 June 2003
12. Sander O, Glas B, Roth C, Becker J, Müller-Glaser K (2009) Design of a vehicle-to-vehicle communication system on reconfigurable hardware. In: International conference on field-programmable technology (FPT), Sydney (to be published)
13. Sander O, Glas B, Roth C, Becker J, Müller-Glaser K (2009) Priority-based packet communication on a bus-shaped structure for FPGA-systems. In: Design automation and test in Europe, Nice
14. Glas B, Sander O, Stuckert V, Müller-Glaser K, Becker J (2009) Car-to-car communication security on reconfigurable hardware. In: 69th vehicular technology conference, Barcelona

15. Sander O, Merz J, Becker J, Reichmann C (2011) Automatic gateway prototype generation for optimization of E/E-architectures based on high-level models. SAE technical paper, SAE 2011 World Congress & Exhibition, Detroit, April 2011
16. AUTOSAR partnership—Fennel et al (2006) Achievements and exploitation of the AUTOSAR development partnership. In: Convergence 2006—Session 1: International Standards
17. Albers A, Dueser T, Sander O, Roth C, Henning J (2010) X-in-the-Loop-Framework für Fahrzeuge, Steuergeräte und Kommunikationssysteme. In: ATZ Elektronik, Ausgabe 05/2010
18. Albers A, Düser T, Ott S (2008) X-in-the-Loop als integrierte Entwicklungsumgebung von komplexen Antriebssystemen. In: 8. Tagung Hardware-in-the-Loop-Simulation Haus der Technik, Kassel

Traffic Signal Information in a Real Residential Area

Benno Schweiger, Regina Glas, Christian Raubitschek and Johann Schlichter

Abstract In this study we share an evaluation of measurements performed in a traffic light communication test bed in real life traffic. We describe our hardware and software architecture and present our measurement methods. As a basis for the evaluation, we selected two use cases: Micropause Infotainment and Fuel Efficient Traffic Light Approach. We develop, train and evaluate a model for estimating micropauses at traffic lights and determine the value of predictive traffic light information in terms of fuel efficiency.

1 Introduction

In the development of modern cars, increasing fuel efficiency and comfort are two of the most important goals. While the need to increase fuel efficiency is mostly driven by legal and financial constraints, the urge for comfort is fundamentally human. To achieve these two goals, car manufacturers increasingly employ prediction based systems. The systems create a prediction about the future state of a vehicle. The prediction is based on data available in the car, like digital maps, traffic sign database and online traffic information. In the near future, another important data source will be available: Traffic light information. Various research

F2012-I02-004

B. Schweiger
BMW Research and Technology, Munich, Germany

R. Glas (✉) · C. Raubitschek
BMW Group, Munich, Germany

J. Schlichter
Technische Universität München, Munich, Germany

projects dealt or are still dealing with this issue: (pre)DriveC2X, simTD, NoW, Aktiv. In these projects, a traffic light periodically sends information about its current and future state using Car2Infrastructure communication. Theoretical studies show a great potential for fuel efficiency (e.g. [1]). These theoretical studies usually act on simplified, optimized assumptions about traffic. To get a feeling for the practical values, we want to measure the effects of traffic light communication in real traffic. The amount of fuel saved depends on two factors: The quality of the prediction and the performance of the system that acts on this information. This paper focuses on the first aspect; the second is only mentioned marginally. We want to measure, how much a prediction can be improved, if traffic light information is available.

For the measurements, we had the possibility to use the test bed ElisaTM, a test bed consisting of several traffic lights in public traffic [2].

The remainder of this paper is structured as follows: After a review of related work, Section "Evaluation Use Cases" lists the use cases towards which the evaluation will be performed.

Section "Measurements" describes the measurement setup. The Evaluation of the results can be found in section "Evaluation". Section "Conclusion and Outlook" concludes this paper and gives an outlook on future work.

2 Related Work

As stated in the introduction, traffic light communication has been the focus of many research projects. Message formats to be used are being standardized as SAE J2735 [3]. Practical experiences have been made in government funded research projects like DRIVE-C2X [4] and Aktiv [5]. In these projects, the focus was on the technical aspect, e.g. how data can be transmitted and which latency times may occur. Tests have mostly been done under artificial conditions in closed test beds. In our work, we profit from the results of these projects, as they offer a sound technical basis for our measurements.

The project simTD [6] is currently developing and constructing a large scale test bed for Car2Car and Car2Infrastructure-Communication application in the area of Frankfurt, Germany. The planned experiments focus mainly on the effect of Car2X-Communication on traffic safety and flow. Our work has a more microscopic focus as we concentrate on the effect traffic light communication has on a single driver and inside the car.

On the theoretical side, much effort has been put into simulating the effect of traffic light communication. To name only a few, [1] and [7] have developed algorithms that use information of upcoming traffic lights to plan a fuel optimal speed profile. Their simulation results show a reduction in fuel consumption of 12–47 %. Since the simulation scenarios assumed optimal conditions, these numbers can be seen as theoretical upper limit that will be hard to realize under realistic circumstances.

In [8] the authors have calculated how much fuel can be saved in a single traffic light approach depending on the distance at which the information becomes available and the cycletime at which the traffic light is when the car receives the information. The results show an average fuel saving of 4 ml at an activation distance of 300 m.

3 Evaluation use Cases

Traffic light information has many possible applications, ranging from simply making a driver feeling informed to safety applications like a red light warning. In this study, we will concentrate on two use cases: Micropause Infotainment and fuel efficient traffic light approach.

3.1 Micropause Infotainment

The goal of Micropause Infotainment is to make short phases of standing still useable for the driver. These pauses may occur at many occasions in traffic, e.g. waiting times for the right of way or small congestions. In [9] the author shows that it is important to know the expected duration of a pause in order to be able to use it optimally. If a pause occurs in front of a red traffic light and the phase changing times of this traffic light are known, the duration of a pause can be estimated.

In this study, we will measure, how much time can be made available to the driver on a drive through inner city traffic regulated by traffic lights.

3.2 Fuel Efficient Approach

An important aspect when talking about a fuel efficient traffic light approach assisted by traffic light information is the distance and time interval between receiving the information and reaching the traffic light. The more time and space is available, the greater is the potential for saving fuel.

In this study we will measure where and when traffic light information becomes available at inner city traffic lights in regular traffic.

4 Measurements

For the measurements, we were able to use the test bed ElisaTM in the city of Munich, Germany [2]. Our test vehicle was a BMW 5-series sedan equipped with the necessary communication module.

Fig. 1 Positions of the roadside stations (Copyright 2010 Google, 2010 Digital Globe, GeoContent, AeroWest, GeoEye) [2]

(a) Intersection 1   (b) Intersection 2

4.1 Test Bed

The test bed consists of four traffic lights, of which for technical reasons only two were useable during our measurements. Detailed information about the test bed can be found in [2].

The two intersections are depicted in Fig. 1. The arrow points to the location of the antenna of the roadside station. Due to the positioning of the antenna, signal reception on the roads leading east–west is very poor because of non line of sight conditions. To avoid these poor conditions, we performed measurements only while driving on the north–south leading road.

4.2 Measurement Vehicle

The measurement vehicle we used was a BMW 5-Series sedan. It uses a Denso Wireless Safety Unit running ACUp [10] as communication device. The WSU was connected to two 5.9 GHz antennas. One was mounted on the highest point of the roof, the other on the front left hand side.

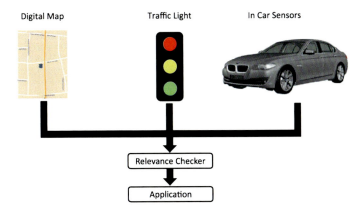

Fig. 2 Components of traffic light software

4.3 Software

The basic components of our software are shown in Fig. 2. Information is taken from three sources: The in-car digital map, the traffic light and the in-car sensors. The traffic lights in the test bed send two types of messages: Signal Phasing and Timing (SPAT) and Intersection Geometry (MAP) messages. The message types are described in [3].

When a MAP message is received, the contained intersection geometry is matched to the digital map and stored in an in-car database. On reception of a SPAT message, a relevance check is started. The first step is to look for a corresponding intersection geometry. If no geometry could be found, the message is discarded, because it is unknown, where the intersection is located. If the geometry is known, it is matched to the planned route of the vehicle. If the match is successful, it returns the signal group that is responsible for the lane and turn direction the ego vehicle will take. For this signal group the current state, timing information and distance to the stop line is handed to the application.

During the course of the measurements, all relevant values were logged with a frequency of 10 Hz. In total, the measurements contained 456 traffic light passes corresponding to a total distance of 320 km.

4.4 Measurement Design

The measurements are taken by driving through the test bed in regular traffic. The drivers were instructed to follow traffic rules and drive as they would normally. To avoid the possibility of a driver getting influenced by the information the traffic lights broadcast, no information at all was given to the driver. The software just logged all necessary data items in the background.

Table 1 Basic sample properties

Total Passes:	456	Number of signal changes:	370
Passes without signal change:	179	Changes red→ green:	130
Passes with signal change:	277	Changes greed → red:	240

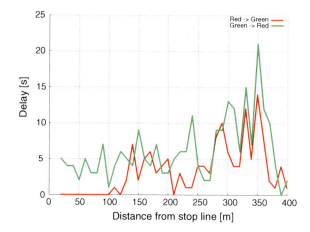

Fig. 3 Distance, at which the information about the next phase change was available

4.5 Results

Measurements were taken during a four month period by different drivers. The total distance travelled in the area of the test bed was approximately 320 km. Table 1 shows basic properties of the sample measured. The sample size of 456 traffic light passes can be considered large enough to create valid conclusions. However, due to the fact, that the test bed consisted of only two traffic lights, the results of this study can only be regarded as single case bound with no claim to universal validity.

The number of phase changes observed is greater than the number of passes with signal change. This is explained by the fact that at an approach, where the light turns from green to red, at least two changes will be observed: The one from green to red, forcing the car to stop in front of the traffic light, and the one from red to green, allowing the car to pass the traffic light.

5 Evaluation

As stated in Subsection "Results", due to the limitation to two traffic lights, the results are not universally applicable. However, the type of intersection we used to perform the measurements is typical for inner city or suburban road networks, making an evaluation using the two use cases mentioned in Section "Evaluation Use Cases" valuable (Figs. 3 and 4).

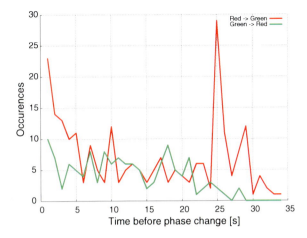

Fig. 4 Time span between information about the next phase change becoming available and the phase change

5.1 Micropause

For the efficient utilization of the short periods of time, when the car is standing still in traffic, it is important, to know, when the pause will end. If the end is unknown to the driver, he cannot devote his full attention to, e.g. reading an email, but has to keep an eye on the traffic.

The total time of standing still at a red light (t_{pause}) consists of two parts: The period until the light turns from red to green (t_{red}) and the period until the backlog in front of the ego vehicle is dissolved and driving can continue (t_{red}) The first period is transmitted by the traffic light, so the ego-vehicle has knowledge of it. The second period is determined by the number of cars in front of the ego-vehicle and the reaction times of their respective drivers $t_{reaction_i}$ (see Eq. 1).

$$t_{pause} = t_{red} + t_{delay} \qquad (1)$$

$$t_{delay} = \sum_{i=1}^{n} t_{reaction_i}$$

Sadly, the number of cars cannot yet be determined by any sensor, neither can the reaction time of the other drivers be determined. However, under the assumption that the size of a single car in a backlog is rather constant, the number of cars can be estimated by the distance of the ego-vehicle to the stop line of the traffic light (d_{ego}). The reaction time of drivers must also be assumed as rather constant. This assumption is backed by studies like [11].

With car length and reaction time assumed as constant, Eq. 1 suggests a linear correlation between t_{delay} and d_{ego} (see Eq. 2).

$$t_{delay} = a * d_{ego} + b \qquad (2)$$

In order to evaluate this hypothesis, we built an estimator for t_{delay} using half the samples, leaving the other half to evaluate the estimator. A linear regression using

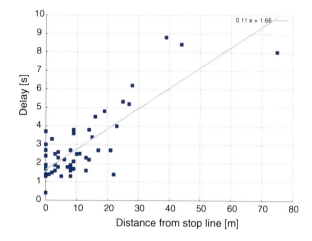

Fig. 5 Delay between traffic light turning green and start of driving

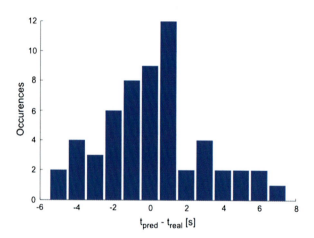

Fig. 6 Estimation error of the linear estimator

the least-squares estimation yields $a = 0.11s/m$ and $b = 1.66s$. These values are in the same range as the ones determined by an empirical study [12]. The relation between d_{ego} and t_{delay} is given in Eq. 3.

$$t_{delay} = 0.11\frac{s}{m} * d_{ego} + 1.66s \qquad (3)$$

Figure 5 shows the training samples and the resulting regression straight line. Considering the circumstances, a Pearson product-moment correlation coefficient of 0.78 strengthens the initial assumption about a linear relation between d_{ego} and t_{delay}.

The results of the cross validation with the other half of the samples not used to train the model are illustrated in Fig. 6. A negative error means the pause is longer than expected. The mean error $\bar{x}$ is -0.29 s, the standard deviation σ 2.72 s.

Traffic Signal Information

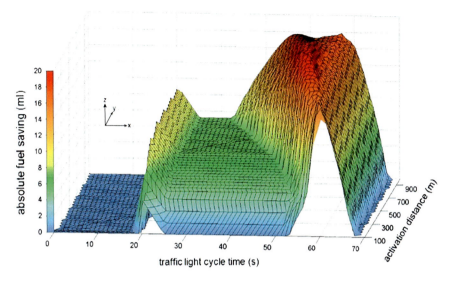

Fig. 7 Simulated fuel consumption reduction in dependence of the distance [8]

According to [9], a micropause application should fade out approximately 3 s before the pause ends. This means that if the pause is predicted 3 s too long, the application will fade out right when the pause ends. While this is not optimal, it is still acceptable because it resembles the behavior of an uninformed driver having no information about the end of the pause. If the pause is longer than predicted, the application fades out before it is necessary. This results in a lower utility for the driver compared to a perfect prediction. But since it doesn't cause any critical traffic situations, an error of -5 s can be considered acceptable. The interval $[-5 \text{ s}:3 \text{ s}]$ contains 84.2 % of all predictions.

5.2 Fuel Efficient Approach

In [8] the authors calculated the amount of fuel that can be saved on passing a single traffic light by adjusting the driving strategy to information about the future state of this traffic light. The amount depends on two influences: The distance, at which the traffic light information is available and the current position in the traffic light cycle. The first influence is caused by the fact that the greater the distance at the point in time when traffic light information becomes available is, the greater is the possibility to react to this information. The second influence determines the kind of possibilities available at that point in time. Because of external constrictions (e.g. maximum allowed speed), the most efficient driving strategy is not always possible. In other cases, the most efficient strategy is also chosen by the uninformed driver: If the light stays green until it is passed, no saving can be realized by additional information. The fuel consumption reduction depending on these two influences is illustrated in Fig. 7.

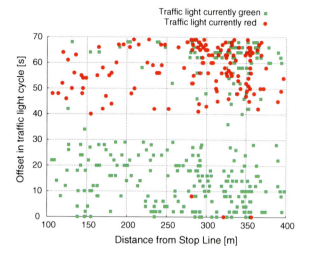

Fig. 8 Distance, when information becomes available and traffic light cycle position

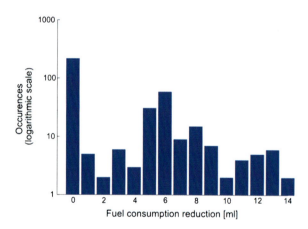

Fig. 9 Distribution of the potential for fuel consumption reduction per pass

Since the absolute amount of fuel saved by predictive information is rather small compared to the noise in fuel consumption caused by traffic and weather conditions, measuring it in real life would have required an enormous number of measurements. So we chose to perform the evaluation based on the data calculated by [8]. For this evaluation, we had to map the data to the format expected in [8]. Figure 8 shows the result of this mapping. Because [8] only regarded activation distances of over 100 m, only measurements with an activation distance greater than 100 m could be used for further evaluation.

The gap of measurements between 30 and 40 s cycle time is explained by the fact, that the traffic lights in the test beds have a shorter red phase than the one assumed in [8]. The distribution of the possible fuel consumption reductions per pass can be seen in Fig. 9.

Table 2 Aggregated fuel consumption reduction

Avg. reduction per pass:	2.04 ml	Total consumption:	32 l
Total reduction:	932 ml	Relative reduction:	2.9 %

The high number of passes without any potential for fuel consumption reduction is caused by the way, the traffic lights in the test bed operate: They give priority to the main road, on which the measurements were performed. This preference results in many passes on green, without the need to deviate from the way an uninformed driver would behave. Table 2 shows the aggregated values for the fuel consumption reduction. Over all 456 passes in the sample, an average of 2.04 ml fuel can be saved by predictive traffic light information, which is equivalent to a relative reduction of 2.9 % in inner city traffic.

6 Conclusion and Outlook

The results, the evaluation of our measurements in a traffic light communication test bed delivered, prove that the expectations raised by previous theoretical studies can be transferred into practice to some degree.

The precision of the micropause prediction suffers from the non predictability that adaptive traffic lights show. Usually, the traffic light controller decides on whether to extend a red or green phase at the end of that phase depending on the current sensor input. Before that decision, the end of the phase can only be estimated by a minimum and a maximum value. Future traffic light controllers might employ a statistic analysis of previous phases to enhance the phase length prediction.

The evaluation of the fuel efficient traffic light approach suffers from not being able to show reductions of this magnitude in a small scale real test. The effect random traffic events or even weather conditions have, drowns the benefit, the predictive information has. Nevertheless, we consider the results of the theoretical evaluation sound enough to merit further research in this area. Future work should focus on how to include other road users (e.g. a backlog at the traffic light) into determining the optimal driving strategy.

References

1. Asadi B, Vahidi A (2010) Predictive cruise control: utilizing upcoming traffic signal information for improving fuel economy and reducing trip time. IEEE Trans Control Syst Technol 19(3):707–714
2. Schweiger B, Raubitschek C, Bäker B, Schlichter J (2011) ElisaTM—car to infrastructure communication in the field. Comput Netw 55(14):3169–3178
3. Society of Automotive Engineers I (2009) Draft SAE J2735 Dedicated short range communications (DSRC) message set dictionary. Elements

4. DRIVE C2X Consortium (2012) DRIVE C2X—accelerate cooperative mobility. From http://www.drive-c2x.eu/. Accessed July 10, 2012
5. Aktiv Forschungsinitative (2012) aktiv Website. From http://www.aktiv-online.org/. Accessed July 10, 2012
6. simTD Consortium (2012) simTD Website. From http://www.simtd.org/. Accessed July 10, 2012
7. Mandava S, Boriboonsomsin K, Barth M (2009) Arterial velocity planning based on traffic signal information under light traffic conditions. In: 12th international IEEE conference on intelligent transportation systems, St. Louis, Mo, pp 160–165
8. Raubitschek C, Schütze N, Kozlov E, Bäker B (2011) Predictive driving strategies under urban conditions for reducing fuel consumption based on vehicle environment information. In: IEEE forum on integrated and sustainable transportation systems 2011. Vienna, Austria
9. Gärtner T (2012) Entwicklung von Strategien zur optimalen Nutzung der Mikropausenassistenz. Diploma Thesis, TU Braunschweig
10. BMW Forschung und Technik (2008) KAS—AKTIV communication unit. White Paper. Accessed from http://www.aktiv-online.org/deutsch/Downloads/BMW_ACUp_Communication_Unit.pdf
11. Triggs TJ, Harris WG (1982) Reaction time of drivers to road stimuli. Human Factors Report, HFR-12(June)
12. Hoffmann G, Nielsen S-M (1994) Beschreibung von Verkehrsabläufen an signalisierten Knotenpunkten. Typo-Druck & Verlagsgesellschaft, Bonn

V2X Communication Technology: Field Experience and Comparative Analysis

Heri Rakouth, Paul Alexander, Andrew Jr. Brown, Walter Kosiak, Masao Fukushima, Lali Ghosh, Chris Hedges, Henry Kong, Sven Kopetzki, Ramesh Siripurapu and Junqiang Shen

Abstract The exploration is built upon Delphi's, Nissan's, Cohda Wireless' and Savari's experiences in Asia, Europe and U.S.A. It describes and derives lessons from all four companies' contributions in projects such as SMARTWAY in Japan, Drive C2X and in Europe, as well as the Connected Vehicle Safety Pilot in the U.S.A. All the above programs were implemented by means of the Dedicated Short Range Communication (DSRC) technology in the SHF spectrum based on the IEEE 802.11p/Wireless Access in Vehicular Environments (WAVE) standard. The study is supplemented with insights regarding complementary technologies such DSRC in the lower UHF frequency band (i.e. 700 MHz) as well as a V2X implementation through the 4G LTE (Long Term Evolution) cellular telecommunication technology. This paper addresses issues regarding the physical layer (PHY) of the DSRC system. The combination of the delay profile caused by multipath propagation along with the motion-based Doppler spread leads to time

---

F2012-I02-005

H. Rakouth (✉) · A. Jr. Brown · W. Kosiak · C. Hedges
Delphi Automotive Systems, Troy, MI, USA

P. Alexander
Cohda Wireless, Adelaide, Australia

M. Fukushima
Nissan Motors, Kanagawa, Japan

L. Ghosh · S. Kopetzki
Delphi Automotive Systems, Germany

H. Kong · J. Shen
Delphi Automotive Systems, Pudong, China

R. Siripurapu
Savari, USA

and frequency dispersion. This limits the number of bytes acceptable for reliable communication or requires a solution at the receiver end. The analysis of the Doppler spread shows that DSRC implemented at 700 MHz is more immune from data packet length issues as opposed to 5 GHz DSRC. On the other hand, 700 MHz DSRC exhibits a much longer delay spread. Thus, guard time interval specified in ASTM E2213-03 cannot be applied as is to 700 MHz DSRC. This paper refers to the German project CoCarX and the Japanese SKY for pedestrian for studying the feasibility a V2X system built on the 4G/LTE technology and its infrastructure. It provides on a vision for an accelerated V2X deployment based on a heterogeneous system. Last, we recommend the ITS stakeholders to carry out extensive research and validation works on DSRC capacity for ensuring a large scale deployment.

Keywords DSRC · Latency · Capacity · Deployment · LTE

1 Introduction

Vehicle to Vehicle (V2V) and Vehicle to Infrastructure (V2I) are collectively known as V2X. They are part of an automotive technology enabling automobiles to communicate in between themselves and with the infrastructure for sharing traffic information or providing situational awareness for congestion mitigation and crash avoidance.

The V2X connectivity is enabled by wireless communication technologies. The V2X connected vehicles vision is extensive and has the following objectives: (1) Each vehicle is continuously connected to all nearby vehicles, (2) Each vehicle is permanently connected to the roadway infrastructure, (3) Though many unforeseen V2X applications are yet emerge, V2X is primarily aimed at improving safety, reducing fatality, easing traffic flow and energy consumption, (4) V2X paves the way for cruise-assisted highways and autonomous driving,

Under the Smartway program, Japan has already started to deploy V2I-based systems since 2009 [1, 2]. For the other leading countries, full V2X deployment primarily driven by V2V is expected to start between 2015 and 2020. It will become a basic mobility requirement in most advanced economies by 2025 [2, 3].

For the above deployment timing to become effective, the V2X technology needs to be fully validated. This paper looks at the current status of the V2X with a survey of key Asian, European and U.S. projects. While millions of miles of field operational test (FOT) are being logged and still counting, and standardization activities are being done at a steady pace globally, some of the limits of the V2X are yet to be uncovered. This paper dives into the basics of propagation in order to understand the challenges of mobile communication, not necessarily accounted for in the standard specification. With the lessons learned, we will look into the related field test results in order to compare the limits of the competing technologies according to safety critical criteria such as range, packet error rate (PER), and latency. While most of the analyses are done around the DSRC technology, we are also exploring how the cellular technology, and in particular 4G-LTE fares with respect to these criteria and to what extent it can contribute to the connected vehicle vision.

2 V2X Communication Technology: Global Status

Over the past decade local and federal governments of leading countries in Asia, Europe, and U.S.A. along with the transportation and telecommunication industries have been funding research and field test programs for promoting V2X. The following sections first provide a brief survey of each region's program and additional insights regarding leading programs.

2.1 V2X Projects Survey

2.1.1 Asia

Many institutes in Asia, such as the Research Center of ITS Engineering in China, Information and Communications Research Laboratories/Industrial Technology Research Institute in Taiwan, Agency for Science, technology and Research/Science and Engineering Research Council in Singapore have been running research and development of V2X communication. Primarily under the Advanced Safety Vehicle program Japan has been working on Advanced Driver Assist Safety (ADAS) and V2I-based applications since 1991. ASV has entered its 5th development phase with the introduction of V2V since 2007 [4]. The other major projects in Japan are Electronic Toll Collection (ETC), Vehicle Information and Communication System (VICS), Advanced Cruise Assist Highway Systems Research Association (AHSRA) and SMARTWAY [2, 5].

2.1.2 Europe

The main projects in Europe are: COMeSafety (Communications for eSafety), SAFESPOT,
 CVIS (Cooperative Vehicle-Infrastructure Systems), SEVECOM (Secure Vehicular Communication), DRIVEC2X, Safe and Intelligent Mobility (sim^{TD}), and Ko-PER (Cooperative Perception). Most of them are funded under the European Union (EU) Framework Programme currently unfolding as FP7 [2, 6].

2.1.3 USA

The main projects involved in the development of V2X communications in the USA. are: IntelliDrive/VII (Vehicle Infrastructure Integration), Vehicle Safety Communications (VSC),

Vehicle Safety Applications (VSC-A), Cooperative Intersection Collision Avoidance System (CICAS), V2V Communication for Safety [2, 6].

2.2 Select Activities

2.2.1 Smartway

Smartway is a nationwide cooperative ITS project led by the Road Bureau of the Ministry of Land, Infrastructure Transportation and Tourism (MLIT) of Japan. It is the continued implementation of the Comprehensive ITS Plan (CITSP) which includes two stages. The first stage started with the implementation of the Vehicle Information and Communication System (VICS) that provides real-time traffic information to car navigation system. The Electronic Toll Collection (ETC.) system begun in 2001 and concluded the first stage. Smartway started in 2007 as the beginning of the ITS in the second stage. Smartway can be defined as a road system enabling exchange of various types of information between cars, drivers, pedestrians, and other roadway users [5] (Fig. 1).

2.2.2 SKY

The SKY Project (Start ITS from Kanagawa, Yokohama) launched in October 2004 in Yokohama, Japan, was aimed at helping the local community reduce traffic accidents and congestion. SKY is a private sector collaboration involving Nissan Motor Co. along with some telecommunication companies. In addition, public sector support was primarily provided by the National Police Agency of Japan (NPA) and the Kanagawa Prefectural Police. SKY is based on 3 services: (1) Dynamic Route Guidance, (2) Safety Drive Support through road conditions, (3) Electronic Toll Collection (ETC) [7] (Fig. 2).

2.2.3 Drive C2X

Drive C2X is a European project pre-validated through PRE-DRIVE C2X (2008–2010) (Preparation for driving implementation and evaluation of C2X communication technology). Drive C2X is a large scale deployment of cooperative

V2X Communication Technology

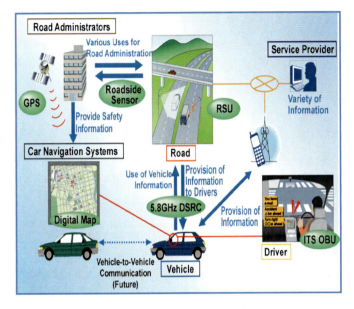

Fig. 1 Overview of the SMARTWAY project

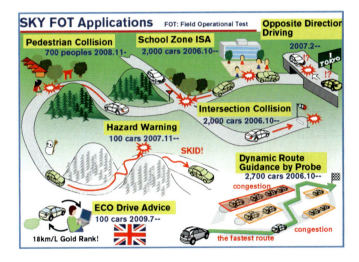

Fig. 2 Overview of the SKY project

systems (IEEE 802.11p DSRC and cellular UMTS radios) over seven test sites in Finland, France, Germany, Italy, Netherlands, Spain and Sweden from 2011 to 2013. It involves major European and Japanese OEMs along with Tier 1 suppliers like Delphi. It will evaluate how ordinary drivers will react to different V2V and V2I services. The main functions evaluated include traffic flow, traffic management, local danger alert, driving assistance, internet access and local information services. Juliussen [2] Delphi's main involvement includes, provisioning DSRC radios, implementing software for facility layer, leading the "FOT framework" part of the project.

2.2.4 KoPER

Ko-PER (2009–2013) stands for Cooperative Perception. It is aimed at capturing a complete picture of the local traffic environment by means of collecting data from distributed sensor networks and subsequently merging them. Sensors can be mobile (vehicle-borne) or stationary (road intersection-based). Each vehicle and nearby roadside unit integrate all relevant data into situational awareness information dubbed Perception that is then shared wirelessly by means of V2V and/or V2I communications for the purpose of data fusion and perception correction (e.g. GNSS positioning). In addition to congestion mitigation, collision avoidance, this project will explore pedestrian's protection by means of "intention recognition" (e.g. "a pedestrian is about to cross the street", "a car is about to turn the corner"). Delphi is the coordinator of the communication working group, provides the distributed sensor fusion center, and develops the algorithm for cooperative relative positioning [8].

2.2.5 Connected Vehicle Safety Pilot

The United States Department of Transportation (USDOT) currently has a very active set of research programs that are focused on the development of crash warning and avoidance systems based on V2X DSRC technology. They wish to accelerate the introduction and commercialization of DSRC-based safety systems. To support this goal, USDOT has initiated a Safety Pilot program starting in 2010 and extending into 2014. It has four major tracks: (1) Vehicle Builds and Driver Clinics, (2) Device Development and Certification, (3) Real-World Testing, (4) Independent Evaluation [9].

The most important of all the stages is the Safety Pilot Model Deployment (SPMD). This will last for a period of 12 months beginning August 21st 2012. The goal is to have 3,000 vehicles fitted with devices that will communicate with other vehicles and roadside infrastructure in a real world scenario. Data collected during the course of these 12 months will be used to evaluate the safety benefits of V2V safety applications in support of the National Highway Traffic Safety Administration (NHTSA) 2013 agency decision for making this a mandatory technology in every vehicle.

The University of Michigan Transportation Research Institute (UMTRI) is the test conductor for the SPMD. The V2X equipment list consists of Vehicle Awareness Device (VAD), Aftermarket Safety Device (ASD), Integrated Safety Systems (ISS), Retrofit Safety Devices (RSD) and Roadside Equipment (RSE).

Vehicle Awareness Devices (formerly known as Here-I–Am (HIA)) only broadcast the Basic Safety Message (BSM) consisting of vehicle's location and dynamics. VAD does not generate warning and is not connected to the in-vehicle data system.

The ASD provides working safety applications. ASD broadcasts BSMs and receives BSMs, traffic light Signal Phase state and Timing (SPaT), map data aka Geometric Intersection Description (GID) and Traveler Information Message (TIM) consisting in road advisory data. ASD includes an HMI. The RSD has the same functionality as the ASD. In addition, it is connected to the vehicle data bus, and thus can broadcast information from in-vehicle sensors. Thus, the RSDs will support safety applications like "Curve Speed Warning", "Forward Collision Warning", "Emergency Electronic Brake Light" etc....The ISS is an RSD that is embedded by the car manufacturer. The SPMD will run over 73 lane-miles of roadway instrumented with 29 RSE installations. For safety pilot purposes, RSEs broadcast SPaT, GID, and TIM. For the Safety Pilot Model Deployment (SPMD), TIM will be materialized as curve speed messages [10].

Savari and Cohda Wireless are the primary radio vendors for the supply of VADs, ASDs and RSEs for the Safety Pilot program.

Delphi provides on-site test support (e.g. configuration of the VADs) in conjunction with Savari. In addition, Delphi supplies the user interface, the HMI framework, and specific V2I applications in cooperation with Cohda Wireless.

3 Technical Perspectives

We will study technical aspects regarding the PHY of both DSRC and cellular technologies.

3.1 Dedicated Short Range Communication Technology

3.1.1 WAVE Standard

WAVE is a set of network standards for vehicular networks operating on the DSRC frequency bands. It supports a major ITS initiative aimed at improving the transportation environment in the aspects of safety, management and data services in a fast environment with less cost. The WAVE protocol stack is built upon the IEEE 1609 family, IEEE 802.11p and SAE J2735. IEEE 802.11p is based upon IEEE 802.11a, and in particular, it provides the Physical Layer (PHY) specifications [11, 12]. It is also worth noting that WAVE's compliance to FCC requirements (i.e. channel and power limits) is ensured through the American Society for Testing and Materials (ASTM) standard E2213-03 (2010) [13]. Because of its reliance on IEEE 802.11a, WAVE leverages Orthogonal Frequency Division

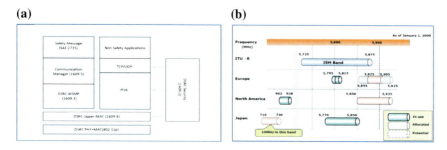

Fig. 3 Overview of WAVE. **a** Protocol stack [28]. **b** DSRC spectrum allocation worldwide [29]

Multiplexing (OFDM) for mitigating the effects of both time- and frequency-selecting fading channel inherent to mobile environments. For V2X implementation, the 5.9 GHz frequency band has been assigned in Europe and North America, and 5.8 GHz in Japan. In addition, Japan has assigned a 700 MHz band initially for research and for effective use from July 2012 onward (Fig. 3).

3.1.2 Practical Requirements for V2X

The following parameters have been experimentally characterized for validating safety applications based on DSRC communications at 5.9 GHz. They are: range, latency, and Packet-Error-Rate (PER). As such, they will be used as references for assessing the performances of other communication modes like DSRC at 700 MHz or cellular-based systems.

Per ASTM, the DSRC range is 1,000 m [13]. But empirical data from the field tests suggest the required range for effective communication is 50–300 m [14]. Latency is defined as the time lag between the time between transmission and reception. Most applications allow a latency of 100 ms but for applications like pre-crash sensing, the allowable latency is about 20 ms [14].

Preliminary analysis showed that the packet size required is approximately 200–500 bytes excluding the security overhead. The size of the security head is approximately 200 bytes [14]. The packet error rate shall be less than 10 % for a message of 1,000 bytes [13].

3.1.3 Channel Metrics

Mobile wireless communications are challenged by multipath components due reflections (see Fig. 4a). They widen the time domain impulse response (see Fig. 4b) and lead to a frequency selectivity of the received spectrum (see Fig. 4c) measured by coherence bandwidth B_c. In addition, a relative motion between

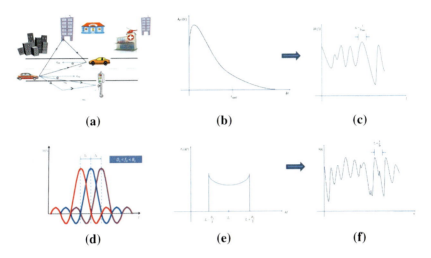

Fig. 4 Overview of mobile wireless challenges. **a** Multipath scenario. **b** Multipath intensity profile. **c** Frequency selectivity → B_c. **d** OFDM subcarrier spacing → f_b. **e** Doppler spread. **f** Time selectivity → T_c

transmitter and receiver induces a Doppler spread (see Fig. 4e) which creates a time selectivity of the time amplitude of the received signal that is measured by the coherence time T_c (see Fig. 4f).

3.1.3.1 Power Delay Profile

Figure 5a shows the Power Delay Profile (PDP) of the wireless channel where each arrow represent the received power $P(\tau)$ as defined by Eq. (1) [15].

$$P(\tau) \approx \overline{k|h_b(t,\tau)|^2} \qquad (1)$$

Where, $h_b(t, \tau)$ is the baseband equivalent channel impulse response. The bar denotes the spatial average over the local area and k represents the channel gain.

3.1.3.2 Delay Dispersion Statistics

The delay dispersion statistics are: (1) the mean excess delay T_{MEAN} representing the delay of the typical echo, (2) the spread of the echoes in time known as RMS delay spread and denoted as T_{RMS} and (3) the maximum excess delay denoted as T_{MAX} that is the time interval between the first and last measurable echo in the channel.

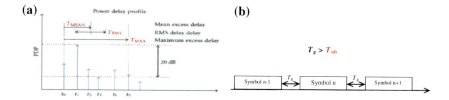

Fig. 5 Power delay profile and its impact on inter-symbol interference. **a** PDP and delay dispersion statistics [18]. **b** Guard time requirement to avoid ISI

Equation 2 defines the "sum delay spread" denoted as T_{sds}.

$$T_{sds} = T_{MEAN} + T_{RMS} \qquad (2)$$

We learn from Fig. 5 that to avoid inter-symbol interference (ISI), two consecutive symbols need to be separated by a guard time T_g greater than T_{sds} That is:

$$T_g > T_{sds} \qquad (2.\text{bis})$$

OFDM is used to effectively reduce ISI. Thus, for an effective fading reduction, the guard time between two consecutive OFDM symbols needs to respect the previous requirement.

The OFDM guard time specified by IEEE802.11.p and ASTM E2213-03 is 1.6 μs.

3.1.3.3 Doppler Spread

Doppler Spread is denoted as D_s. It quantifies the spectral broadening around the carrier frequency f_c caused by the time rate of change of the mobile wireless channel.

Assuming a moving receiver with a relative velocity v_{rel} and an angle of arrival θ, D_s is the total span of the frequency shift around the carrier frequency f_c and is equated by:

$$D_s = \frac{2 f_c v_{rel} \cos \theta}{c} \qquad (3)$$

Under OFDM, ICI is efficiently addressed when:

$$Ds < fb \qquad (4)$$

3.1.3.4 Coherence Bandwidth

The coherence bandwidth denoted as B_c is a statistical measure of the frequency band over which the channel is frequency invariant. The OFDM equalization process consists in estimating the channel response by a training sequence transmitted at the beginning of each packet. This results in a set of channel estimates for each subcarrier. In order to achieve a reliable estimate over the entire band, the channel response should be relatively flat between each subcarrier. Consequently, the value of B_c needs to be larger than the subcarrier spacing f_b. That is: $f_b < B_c$. Combined to (4), this last condition leads to:

$$D_s < f_b < B_c \qquad (5)$$

3.1.3.5 Coherence Time

The coherence time T_C, is a statistical measure of the time interval over which, the channel impulse response is essentially invariant, and it is defined as the time duration over which the correlation function is above 0.5 [16]. For an error-free transmission, each data packet shall be transmitted over a time T_T smaller than the coherence time T_c. That is:

$$T_T < T_c \qquad (6)$$

In factoring the expression of T_c as defined in [16], the data rate d_r, and the expression of the Doppler spread D_s for an angle of arrival $\theta = 0$, one can define the maximum packet length (or maximum PHY payload in number of bytes) (denoted as PL_{max}) for an error-free transmission as:

$$PL_{\max}(bytes) \approx \frac{1.502 \times 10^3 \times dr(Mbps)}{f_c(GHz) \times \vartheta_{rel}(mph)} \qquad (7)$$

Equation (7) shows that at 60 mph and for a data rate of 3 Mbps, $PL_{max@5.9GHz} = 127$ bytes and $PL_{max@700MHz} = 1{,}073$ bytes. Thus, DSRC 5.9 GHz will not meet Sect. 8.10.6 of ASTM E2213-03 that requires a maximum PHY payload of 1,000 bytes for 3, 6, and 12 Mbps.

3.1.4 Field Tests

The field test results reported below are aimed to flag potential issues relating to ISI, ICI, or packet length for both 5.9 GHz and 700 MHz DSRC.

3.1.4.1 Maximum PHY Payload Validation for 5 GHz DSRC/WAVE

Over the 2007–2010 period Cohda Wireless has conducted FOTs to validate a proprietary algorithm featuring iterative channel estimates based on outputs of the receiver data decoder [17]. The results are displayed in Fig. 6 and the key findings are:

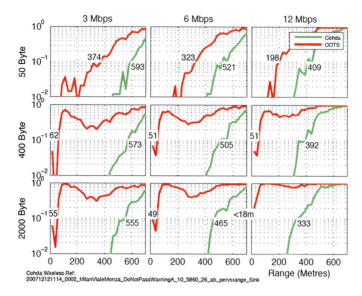

Fig. 6 Comparative range and payload for one-time vs. iterative channel estimate [17]

(a) The COTS radio only achieves the required 10 % PER for the 50-byte payload over a range of 200 m or less for all 3 mandatory data rates. This is in-line with our analysis in Sect. 3.1.3.5.
(b) The Cohda MK1 radio achieves the required 10 % PER for all 3 payloads over a range of 400 m or less for all 3 mandatory data rates.

3.1.4.2 Channel Characterization in the Lower UHF Range

This section reports a channel measurement campaign at 700 MHz conducted in and around downtown Ann Arbor, Michigan, U.S.A. by a research team U.C. Berkeley [18]. (1) The range for the sum delay spread is $T_{sds} < 2.3$ μs. The upper limit is well above the 1.6 μs guard time specified by IEEE802.11.p and ASTM E2213-03. Thus, the 700 MHz channel will be prone to ISI unless T_g is adjusted with a value >2.3 μs. (2) The Doppler spread is $D_s < 63$ Hz and the coherence bandwidth is $B_c > 1$ MHz. Since $f_b = 156.25$ kHz, condition (5) is satisfied. Thus, the 700 MHz DSRC is deemed to be ICI-free. (3) 700 MHz applied to Eq. (7) shows that the 1000-byte packet length will be met.

Since July 2012 and after a 3-year validation period, the 700 MHz band is available for effective ITS utilization and in particular for V2X at blind intersections in Japan [19].

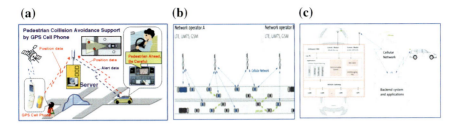

Fig. 7 Two Cellular-based Implementations of V2X. **a** SKY pedestrian safety support. **b** CoCarX system architecture. **c** CoCarX heterogeneous system

3.2 Cellular Communication as a Complement to DSRC

In this section, we will analyze the cellular technology 4G-LTE's ability to fulfill V2X's operations. Then we will describe both SKY and CoCarX implementations (see Fig. 7).

3.2.1 LTE Performance Requirements Versus V2X Practical Requirements

According to its performance specifications [20–22], the cellular technology 4G-LTE is apt at meeting the practical and standard/regulation requirements for V2X. (1) 4G-LTE's 5 km coverage is compliant with the 1,000 m requirement. (2) Its latency of 20–100 ms is well in line with the 100 ms required for most applications. (3) LTE's data rate is 50 for upload and 100 Mbps for upload with a carrier frequency in the 700 MHz, 2 GHz, and 2.6 GHz bands factored in Eq. (7) leads a maximum payload of 2,500 bytes. (4) LTE's guard time of 4.7 μs or more makes it ISI-free.

3.2.2 Overview of SKY Pedestrian Collision Avoidance

Nissan Motor Co., Ltd. and NTT DoCoMo, Inc. conducted a FOT using an app running 3G W-CDMA cellular phones for pedestrian protection November 2008 through January 2009 in a small town near Yokohama city. A local server gathered and calculated the relative position; speed and direction of both the pedestrian(s) and vehicle(s) carrying the app-equipped cell phones (see Fig. 7a). The pedestrian's movement data were sent to the car navigation system which alerted the driver based speed-related safety threshold. 500 pedestrians along with 200 drivers participated in the FOT with a total driving time of 284 h and 4,123 km driven. Test participants received 5 alerts on the average, and 71 % adjusted their driving speed accordingly. An LTE implementation of this application is currently under discussions. General Motors has recently announced a similar project using Wi-Fi Direct [23].

3.2.3 CoCarX

Cooperative Cars Extended (CoCarX) conducted between 2009 and 2011 is a follow-on to the CoCar (Cooperative Cars) is a project run by German car manufacturers along with Vodafone and Ericsson for the purpose of studying the applicability of the cellular technology to ITS including V2X over the 2006–2009 period [24]. CoCarX was primarily focused on evaluating LTE's performance as a standalone communication system or combined with WAVE with cellular (Fig. 8b and c) [25, 26].

CoCarX demonstrated that LTE meets the practical requirement of 100 ms latency for most single pair car-to-car connectivity applications. Latency values between 30 and 100 ms were typically observed under urban conditions. That is the reason why, safety applications such as emergency electronic brake light (EEBL) has been successfully demonstrated. LTE's downlink capacity limits its ability for handling permanent broadcasting for Cooperative Awareness Messages (CAM) such as BSM at 10 Hz. A slower transmission rate at 2 Hz and/or a lesser number of broadcasting cars is more capacity friendly (see Fig. 8a). For example for a 5 MHz spectrum, only 10 vehicles per cell can receive the CAM data sent by 40 neighboring vehicles transmitting simultaneously at a 100 ms latencyNetwork simulation shows that LTE work better for event-driven messages such as DENM (Decentralized Environmental Notification Message). Road Hazard Warning (RHW) and EEBL are two examples of DENM, where the communication is (1) triggered by an authorized event (e.g. presence of road hazards or hard breaking) and (2) broadcast continuously until expiry of the event. As seen in Fig. 8b, DENM can be handled with a reasonable latency of 200 ms or less in typical load scenarios (e.g. 300–600 cars per cells).

At this time no such capacity study is available the DSRC system, but similar capacity issues are expected as pointed out by theoretical analyses [27]. The U.S. Safety Pilot Model Deployment will help document them. In the meantime, more works are needed to address DSRC capacity issues as required

4 Vision for an Accelerated V2X Deployment

The need for permanent broadcasting (e.g. CAM or BSM) along with the requirement for low latency will result in capacity issues in large-scale applications solely based on DSRC. In addition, security functionalities require the installation of RSEs, thus leading to infrastructure development and funding problems that may delay or limit full V2X deployment. As seen in previous sections, such a V2X deployment could be accelerated through a mixed system.

V2X Communication Technology

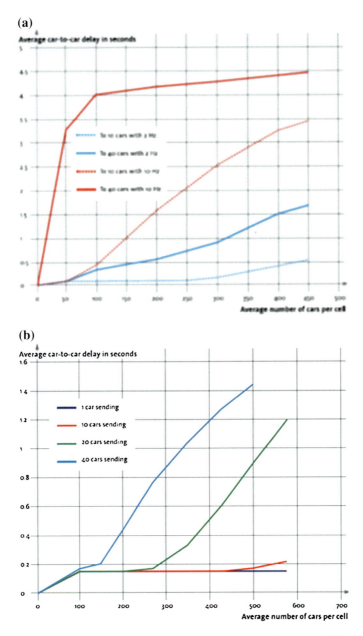

Fig. 8 LTE's latency and capacity performance. **a** CAM Performance [26]. **b** DENM performance [26]

A possible scenario would involve using LTE and or its progeny (e.g. LTE-Advanced with peer-to-peer capability) technology in highway and rural areas for covering most V2X applications including safety applications such as DENM,

RHW, EEBL or in-vehicle signage. With such a cooperative system, 5 GHz or 700 MHz DSRC will be mainly used in congestion and accident prone areas such as road intersections and downtown areas for collision avoidance and pedestrian safety.

5 Conclusion

We reviewed WAVE-based and cellular-based V2X programs in Asia, Europe and USA. The analysis of CoCarX shows that WAVE and LTE can be used cooperatively.

Through theoretical analyses backed by field test results, we have identified the respective strengths and weaknesses of 5 GHz and 700 MHz DSRC.

For 5 GHz DSRC, the practitioners need to be aware that in order to avoid uncontrolled degradation of the PER due to excessive Doppler spread at lower data rate (e.g. 3 Mbps) the maximum packet length should be limited to 100 bytes or less. Otherwise, advanced receivers will be required for longer packets.

Though endowed with excellent range, coverage, and penetration properties, making it a good choice for road intersection V2X, 700 MHz DSRC requires a guard time of 3 μs or higher.

LTE is immune from the above described issues. However, due to identified capacity issues, in its current, this technology may fall short of meeting the latency criteria in congested area for CAM and BSM messaging. Instead, LTE will be competitive for event-driven communications (e.g. EEBL and in-vehicle signage).

As a result, a V2X system using LTE for rural and highway areas along with 5 GHz and/or 700 MHz DSRC would provide a cost-effective mix for accelerating its deployment that will ensure among others, functions such as vehicle crash avoidance and pedestrian protection.

More works are needed for addressing potential DSRC capacity issues.

References

1. Sakanaka Y (2010) ITS radio systems in Japan [P]. ETSI, 2nd ETSI TC ITS Workshop, 10–12 February 2010, France
2. Juliussen E (2012) V2X Technology's arrival key to accident reduction and prevention" [R]. iSuppli, Q2 2012, Topical report
3. Juliussen E, Carlson J, "Automotive research: V2X's current market status and scenarios for future deployment" [R]. iSuppli, Q4 2010, Topical report/automotive research
4. Wani K (2012) "How innovation is driving safety?" [P], Mimistry of land, infrastructure, transport and tourism (MLIT), Japan
5. Makino H, "Smartway project" [P], 12th ITS world congress, November 6–10, 2005. San Francisco, California
6. Karagiannis G et al. (2011) "Vehicular networking: a survey and tutorial on requirements, architectures, challenges, standards and solutions", IEEE communicationns surveys & tutorials, 1553-877X/11/$25.00 c 2011 IEEE

7. Fukushima M, "FOT methodologies and data handing—progress of SKY project" [P], FOT-net second international workshop, Sockholm, Sweden, 21 Sept 2009
8. http://ko-fas.de/english/ko-per—cooperative-perception.html
9. Pina M et al. (2012) "Transforming transportation through connectivity: ITS strategic research plan, 2010–2014, Progress Update 2012" [B], U.S. Department of transportation/research and innovative technology administration/intelligent transportation systems joint program office, technical report FHWA-JPO-12-019, May 2012
10. Sayer J, "Connected vehicle safety pilot", http://www.michigan.gov/documents/msp/SPMD_overview_031912_381173_7.pdf
11. Weigle M (2010) "Standards: WAVE/DSRC/802.11p" [P], Old Dominion University, Spring 2010, CS 795/895
12. Muller M (2009) "WLAN 802.11p measurements for vehicle to vehicle (V2V) DSRC" [AN], Rhode & Schwarz, Application Note 09.2009-1 MA152_Oe
13. ASTM, "Standard specification for telecommunications and information exchange between roadside and vehicle systems—5 GHz band dedicated short range communications (DSRC) medium access control (MAC) and physical layer (PHY) specifications" [S], ASTM, E2213-03, 2010
14. Shulman M, Deering R "Vehicle safety communications in the United States", NHTSA/DOT, Paper number 07-0010
15. Rappaport T (2002) "Wireless communications: principle and practice" [B], 2nd edn. Prentice Hall, New Jersey, p 185
16. Rappaport T (2002) "Wireless communications: principle and practice" [B], 2nd edn. Prentice Hall, New Jersey, p 204
17. Alexander P et al. (2011) "Cooperative intelligent transport systems: 5.9 GHz field trials" [J]. IEEE, 2011, Proceedings of the IEEE, 0018-9219/$26.00 2011 IEEE
18. Sevlian R et al. (2010) "Channel characterization at 700 MHz DSRC vehicular communication" [J], Hindawi Publishing Corporation, Article ID 840895, p 3
19. Yamamoto T (2011) "Activities on advanced ITS radiocommunications" [P]. ARIB, GSC-16 Halifax Canada
20. Chadchan SM, Akki CB (2010) "3GPP LTE/SAE: an overview" [J]. Int J Comp Electr Eng 2(5):806–814. Oct 2010
21. www.nokiasiemensnetworks.com (2011) "LTE Transport Requirements—LTE-capable transport: a quality user experience demands an end-to-end approach" [P], white paper 12 Dec 2012
22. Ghosh A et al (2010) "Fundamentals of LTE" [B], Prentice Hall, New Jersey, pp 21–24
23. General Motors (2012) "GM developing wireless pedestrian detection technology" [J], GM Neews, 26 July 2012
24. Dietz U (2009) "CoCar feasibility study: technology, business and diddemination" [R], CoCar consortium, Pulic report
25. ETSI (2012) "Intelligent transport systems (ITS); framework for public mobile networks in cooperative ITS (C-ITS)", [R], ETSI, technical report ETSI TR 102 962 v1.1.1, Feb 2012, pp 24–33
26. www.aktiv-online.org, "Cooperative Cars eXtended—CoCarX—activating mobile traffic channels" [R], http://amicale-citroen.de/wp-content/uploads/2011/08/CoCarX-Cooperative-Cars-Extended-Research-Project.pdf
27. Li Y (2010) "An overview of the DSRC/WAVE technology" [P], NICTA, Australia, 2010. www.nicta.com.au/pub?doc=4390
28. CAMP Vehicle Safety Communications 2 (2011) "Vehicle safety communications-applications VSC-A: second annual report—January 1, 2008 through December 31, 2008" [R]. U.S. Department of transportation/national highway traffic safety administration, report DOT HS 811 466, August 2011
29. Oyama S, "Activities on ITS radiocommunications standards in ITU-R and in Japan" [P], Hitachi, 2008, 1st ETSI TC-ITS Workshop, 6 February 2009, Sophia Antipolis, France

Improved System for ParkNet Mobile Network

Yao Ge, Wenzhi Xue and Zhan Shu

Abstract In the process of developing ParkNet system, there are several problems limit the accuracy of detection algorithm. Specially, when the sensor vehicle is driving in the two lane road situation, it will greatly influence object recognition. In the paper, we will analyze some special features of this problem. Also, we will post some parking statistics for the valid parking position data collection. Including, identifying legitimate parking space and location matching algorithm.

Keywords Mobile network · Two-lane road · Populating database · Hungarian algorithm · Visualization UI

1 Introduction

In the ParkNet system, the detection of parked vehicles is based on the sequence of distance sensor reading which collected by an ultrasonic sensor on the side of the testing car. The dip is simply a feature in that sensor reading that caused by a parked vehicle. But the assumption is based on that the car which caused the dip is a parked vehicle. We realized some significant problem will occur when another vehicle is passing by the testing vehicle on the side which the ultrasonic sensor is

---

F2012-I02-006

Y. Ge (✉) · W. Xue
Rutgers University, Newark, NJ, USA

Z. Shu
New Jersey Institute of Technology, Newark, NJ, USA

put, specially, driving in a two-lane road (including two driving lanes in one direction and one parking lane). Also, after going through the data which we have already collected for ParkNet system testing, we find that the system actually has no ability to tell the difference between the dip which is whether caused by a passing car or a parked vehicle. In other word, a false detection might be made which flag a vacuum parking slot as an occupied one. This will greatly decrease the accuracy of detection algorithm. Now if we can find out whether our testing car is driving on the left lane (the sensor is on the right side of the car), then we may know there will be a high chance that a detected vehicle could be a car which is passing by but not a parking car. Based on this requirement, we will try to improve the system from following two main aspects without causing change the existing system hardware and software structure.

Recognition of the two-lane road: Because the similarity of the dip which caused by passing vehicles and parked vehicles. Our analysis will dig out some unique features of passing vehicles.

Identifying legitimate parking space: Since the ParkNet ultimately seeks to direct drivers towards valid, available parking spaces, it is necessary to know where valid spaces locating. The result of detection algorithm is the coordination position of every specific vehicle on the road. But the main purpose of the ParkNet system is to give the user some detail parking information, not just telling you how many cars there are on the roadside. So another crucial information about parking position is needed to generate the whole map view of which parking spot is empty or not. However, because the lack of the data about roadside parking position, we need to process and generate the coordinate of every parking spot in our coverage area. It will be a huge project to hardcode every spot by measuring the GPS data in every parking spot. The purpose of our ParkNet system is about low cost. So, in the rest of the paper, we will also discuss an approach to populating a database with the locations of valid parking spaces which will be able to identify the coordinate of every specific parking slot and automatically create the parking position database by using the data we collected for detecting road-side vehicle. It will greatly reduce our cost of developing the system.

In Sect. 2, some background information and literature review about the ParkNet mobile system will be introduced. In Sect. 2, we will analysis a two lane case problem with discussing features and possible solution to solve it. In Sect. 3, we will present an approach to populating a database with the locations of valid parking spaces. Also, we will talk about the matching algorithm to match the detected parked vehicles with the known locations of valid parking space. Also, a statistical analysis will be conducted to show the parking availability variation trend over the certain period. Finally, the conclusion will be given the in Sect. 4.

2 Background

The traffic congestion in urban area has been a big concern of modern society for long time. Also, the most suffering thing for people in the city is to find a parking position close to their destination, even sometimes it might be an endless loop for a driver. According to a recent study [1, 2], up to 45 % of the traffic in Manhattan is due to automobiles circling the block to look for parking positions. Several projects are working on collecting traffic congestion information, through analysis of the information, a trip planning or route guiding system can help reducing the traffic. However, most of these projects use fixed sensors to monitor road-side parking space [3, 4], thus, the cost of sensors installation for every parking meter or the entrance of the parking lot is a huge budget burden to government or company. Besides, the limited coverage of this static sensor network is not sufficient to generate a whole view of traffic congestion, which makes the collected information being less useful. Such a project running in San Francisco which names SF-Park [5] aims to cover only 25 % of the street parking slots, but it will cost about million dollars. Another drawback of system is it needs continuous wireless connection to transform the data from sensor nodes back central server. Because of the diversity of these parking slot wildly spread, it will need some extra expense to build and maintain the network. In sum, it is not an economical and efficient way to use fixed sensors to detect and monitor the traffic in city.

ParkNet [1] system is a mobile system comprising vehicles installed with sensors collecting parking spaces occupancy information while moving around. It consists of one ultrasonic sensor (in Fig. 1) that simply reports the distance to the nearest obstacle which is on the right side of the vehicle and one GPS receiver records the corresponding location, and a node act as a server to collect and store the data from the GPS and sensors. When the sensor vehicle is moving, the sensor will keep reading and sending data to the server node, the flow of data will be first stored in server, and then the node will transform the stored data to some remote server through wireless network. Also, there is a central server which is responsible for aggregating and processing the data. After the data coming into the server, the analysis algorithm will process and filter the data to generate a sequence of detected car on the specific road. After that, the system will match every parking slot to the detected car and build a visualized map of the road-side parking availability. ParkNet also be able to provide this information to any clients who query the system for searching parking space. The advantages of this mobile sensor approach are obvious: First, whereas one stationary sensor can monitor only one parking space, one ParkNet sensor could monitor multiple parking spaces as the equipped vehicle driving around. Second, the hardware requirement for ParkNet can also be deployed at a relatively low cost. The ultrasonic rangefinder and GPS can be integrated with a laptop PC onboard the vehicle, which provides a time-stamp for each sensor reading.

The correctness of this detection system cannot be verified only base on processing the sensor reading data. In order to achieve this, a webcam is installed just

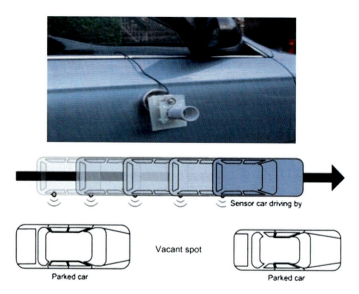

Fig. 1 Drive-by parking monitoring

Fig. 2 A screenshot of groundtruth UI

beside the sensor to capture the image about 20 fps and tags each image with a kernel time stamp. It takes pictures of the real roadside sight while the sensor is working. All the images will be stored into the server database and will be automatically matched to the vehicle which we have already detected from the previous step. Then in server side, a UI is developed to control and manually inspected each image and entered the 'ground truth' information. In our test phase, we can look into some detail data part easy through the interface. An example picture is presenting as in Fig. 2.

Improved System for ParkNet Mobile Network

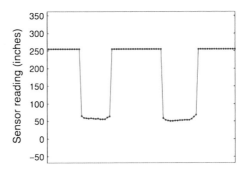

Fig. 3 Sample of the sequence of the sensor reading

Prototype hardware was developed for experimentation purposes. The model consisted of a low-cost ultrasonic rangefinder, a PlayStation 3 Eye Webcam, a Garmin 18-5 Hz GPS and a lightweight computer onboard the vehicle. The webcam was used only for research purposes, to manually evaluate the accuracy of ParkNet's detection algorithms. The GPS provides latitude and longitude coordinates for each sensor reading, and the onboard computer is used to store the data. The rangefinder computed distances at a rate of 20 measurements per second (20 Hz). For each new sensor reading, a string of data is uploaded to the server which appears in the following format: <Time stamp, sensor range, latitude, longitude, vehicle speed>. This prototype hardware was deployed to gather data during a 2-month experiment in Highland Park, New Jersey. Roughly 500 miles of data were collected over a street segment containing 57 slotted parking spaces.

Detection Algorithm After the data uploaded to a server, an algorithm must be executed to detect parking vehicles. The sensor output and time stamp are used to compute the distance to the nearest object as a function of time. Notice that two dips showing in a sample result (Fig. 3) indicate where vehicles have parked. The detection algorithm factors into account the moving vehicle's speed, the dip length, and the dip magnitude (depth) to determine if a dip is in fact a parked vehicle. The algorithm has approximately 95 % accuracy. The 5 % failure rate is mostly attributable to false positives: objects such as flower pots and bicyclists may wrongly appear as parked cars to the algorithm. The detection algorithm translates the sequence of the sensor reading into the coordinates of parked vehicle on the road side. The original data from two equipments (ultrasonic sensor and GPS) need to be processed first. Because the different source of input data, these two data flows need to be merged first. Another problem is, the processing rate of ultrasonic sensors is about 20 times per second, but GPS is about one time per second. Therefore, after merger, there will be some gap between every column of sensor reading. The linear square estimation based interpolation will be implemented to fill these intervals. After initial data processing, we will have a data flow which including timestamp, coordinate, sensor reading, speed and distance. The distance sensor reading can provides a one-dimensional view of distance to the nearest obstacle when the sensing vehicle moves forward.

3 Indentifying Legitimate Parking Space

Fragment of Statistics The ultimate goal of ParkNet is to direct drivers to the locations of nearby, vacant parking spaces. This goal is made difficult by the fragmentation of information about the location of valid parking spaces. The availability of such information varies between cities, and oftentimes, is inaccessible or nonexistent. A scalable model will require a comprehensive database including the coordinates of all relevant parking spaces. Thus we create such a database and populate it with information gathered during our experiment.

Dynamic of Parking Laws Parking laws vary significantly between locations. Some spaces only allow occupancy within a certain time frame, for example, between 8 a.m. and 4 p.m. Others may impose a limit for the duration of time a single car may be parked there, such as 2 h. There are also paid and unpaid spots, and even those prices vary as a function of time. A space which requires an hourly fee during the week may become free on a given Sunday. The variability of parking laws seems innate, which makes constructing a database to manage all of this information difficult. Just like the fragmentation of location statistics described in Sect. 2, statistics about the governing laws of parking spaces are also fragmented. Thus our approach to solving this problem is similar. By implementing a database that populates itself as sensor vehicles gather data, the system will become more intelligent about specific parking laws as they apply to specific locations. For instance, if parked cars are only detected at given location between 8 a.m. and 4 p.m., it is likely that parking is only allowed within that time frame.

Populating a Database of Valid Parking Spaces Since the ultimate goal of ParkNet is to direct drivers towards nearby parking spaces, thus it is necessary to have a database of where parking spaces exist. We do this by compiling the data from roaming sensors and populating a database with information about where cars have parked in the past. For example, some location which never has a parked car most likely is not a valid parking space. Conversely, a location which occasionally has a car parked there probably is a valid space. Thus by storing the locations of all places where vehicles have parked, this database can estimate what is most likely a valid parking space. As more data acquired, the database becomes more comprehensive and accurate, and thus scales well.

Because of memory restrictions, it is unfeasible to store every coordinate where sensors have traveled. To prevent excessive data collection, our database stores only the locations at which the detection algorithm has returned true, namely, locations where cars have parked. Since the database stores only the number of times a vehicle has parked in a given location, estimating space validity by the percentage of times a car has been parked there is impossible. Our alternative approach is to employ an absolute threshold: if a vehicle has been parked in a given location 4 times, assume it is a valid space and augment that space to the database of valid locations.

After populating the database with data acquired from the November 2009 experiment, this approach proved to be partially successful. However, less than

50 % of valid parking spaces were correctly identied. One possible explanation is that such an approach depends on the quantity of data: the more data acquired from a particular location, the more likely that 4 cars will park there at some point in time. A time span of 31 days may not be long enough to successfully identify all valid parking spaces. One advantage of this approach that was made evident in the experiment was the apparent under-approximation of valid parking spaces. Since the algorithm only deems a "space\valid" after having been occupied 4 times, it is unlikely that a location will be "mislabeled\valid". However, it is much more likely that a space will be mislabeled\invalid", because not enough data may exist. Thus ParkNet will rarely direct drivers towards parking spaces which do not exist.

Spots Matching The previous section assumes that ParkNet successfully associates a detected car to the correct space in a database. In reality, this task is completely non-trivial. First, it is extremely unlikely that 2 vehicles detected in the same parking space will actually return the same GPS coordination. The error of the GPS and the variability in parking style prevent us from simply denying equality: if the coordinate of A equals the coordinate of B, then cars A and B are parked in the same location. This situation is obviously true, but incomplete. If the coordinate of A is very close to the coordinate of B, cars A and B may in fact be parked at the same location. Evaluating this situation is what we call\spot matching".

Computing the Cardinal Direction: While attempting to spot-match, we constantly needed the vehicle's heading. The GPS is often precise enough to differentiate which side of the road a vehicle is on, which is necessary to determine which side parking lies on. In our experiment, a magnetometer or other compass-device was not used, so our approach was to determine the heading of a vehicle based on the change in its GPS coordinates.

Computing a vehicle's heading can be done with simple trigonometry as shown in Fig. 4. The first pair of coordinates, A, in a vehicle's string of coordinates is selected. The algorithm iterates through the remaining coordinates until detecting one that is approximately 25 m from the original, B. Let A be the origin of a Cartesian plane and map B appropriately.

The displacement vector between A and B makes an angle Φ with the horizontal, which is to be: $\phi = \arctan \frac{\Delta Long}{\Delta Lat}$. However heading is typically reported as the angle made clockwise from north. In the first quadrant, then, our heading is defined to be 90 $-\Phi$. We account for the second, third and fourth quadrants, by analyzing the signs ($\pm$) of Lat and Long.

GPS Offset Correction Our pilot studies revealed a discrepancy between the true location of a parked space and the GPS reading. This discrepancy is attributable to a number of factors: First, the marginal error of the GPS reading reduces the accuracy of our data. Second, the location of the GPS is inherently separate from the location of a perceived parked vehicle, because the GPS is attached to the mobile vehicle, not the parked vehicle. For instance, a mobile vehicle traveling northeast on Route 27 in Highland Park, New Jersey, may return the GPS coordinates 40.300057, 74.253021. However the ultrasonic sensor may detect a parked

Fig. 4

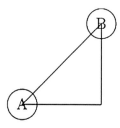

Fig. 5

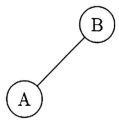

vehicle located at coordinates 40.200040, 74.2530215, approximately 3 m from the location of the GPS. Thus the data ultimately stored in the database for the location of parked car will be 3 m from the true location of that car.

Although this discrepancy may seem trivial, it leads to substantial problems when attempting to match new data about vacancies to the reference map of known parking spots. The discrepancy, along with the random GPS error, often causes our spot-matching algorithm to incorrectly link a vacancy on one street with a known parking space on another street. The problem manifests itself near intersections, where the coordinates of parking spaces on another street tend to be very close.

Mathur et al. [1] proposed an "environmental fingerprinting" approach to solving this problem. However, this relied on monitoring stationary objects on roads, and thus required storing more data and performing more computation. We have developed a new approach to correct GPS o set that relies on the existing reference map of known parking spaces. The first occupied space in a string of spaces is selected. The distance and direction between that selected parking space and every other occupied parking space on the road are then calculated. This process is repeated for each parking space within the given segment of road.

This algorithm then generates a graph, comparing the distances from each occupied parking space to the next. The offset is then corrected by comparing this graph to a graph of the known locations of valid parking spots, or the ground truth.

Figure 5 represents a graph of two detected cars, A and B. Figure 6 represents the corresponding reference map, but there is an additional space, C. This means there is one unoccupied space. The offset correction algorithm computes the distance between A5 and B5. It then computes the distance between A6, B6, and C6. It then notices that the path length between A5 and B5 is nearly identical to that between A6 and B6. If the discrepancy between these 2 displacements is less than

Fig. 6

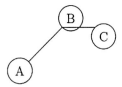

3 cm, we can correct the offset. This offset correction algorithm proves to be extremely accurate. Below are two images comparing the locations of detected cars to the locations of known parking spaces, before and after the algorithm was executed.

In Fig. 7, the points on the upper line describe the locations of detected cars. The points on the lower line describe the actual locations of valid parking spaces. Notice a significant gap between the lines. However in Fig. 8, after the offset has been corrected, notice that the gap decreases substantially. In fact, the offset correction algorithm successfully reduces the gap to less than 3 cm in each case.

Hungarian Algorithm: The offset correction algorithm functions by comparing graphs to reduce the discrepancy between perceived locations and ground truth. It does not, however, uniquely link one detected vehicle to one known parking space. To accomplish this, we have devised an application of the Hungarian Algorithm, developed by Harold Kuhn in 1955, to best optimize the "cost" of our function [6]. "Cost" means the sum of all of the distances between detected cars and their respective parking space. This is done by creating a matrix, and then minimizing the total cost of the matrix.

We will create a matrix C, in which the columns are the locations of known parking spots in the reference map (r). The rows are the locations of detected vehicles (d). Each element in the matrix, $X_{d,r}$, is the distance between the location of a detected vehicle and the corresponding reference spot (Fig. 9).

To successfully match detected cars to their correct reference spots, our algorithm finds the least-cost solution to the matrix. Every column in the matrix can only provide 1 element to the total cost, because only 1 car can fit in a parking space. Likewise each row can only provide 1 element to the total cost, because a detected vehicle can only exist in 1 parking space. So, the least-cost solution to the matrix is that which has the smallest sum of all the highlighted elements.

Development of User Interface The ultimate goal of ParkNet is to provide drivers with information about nearby parking availability through some front-end user interface. Possibilities include a website, mobile phone application, or in-car GPS application. To fully test the capabilities of our back-end infrastructure, we have built a pilot front-end website. Feel free to test the site, which is publicly available, at: http://orbit-lab.org/parknet/index.jsp?current=true?.

The site, however, is frequently inaccessible or dysfunctional because of our ongoing research. Also, we have developed a web interface for the database, which includes photos of all detected parked vehicles. The site's URL is: http://orbit-lab.org/parknet/database.jsp.

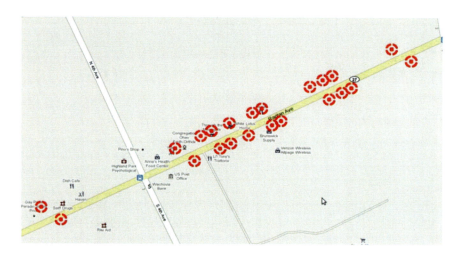

Fig. 7 Map view before offset correction

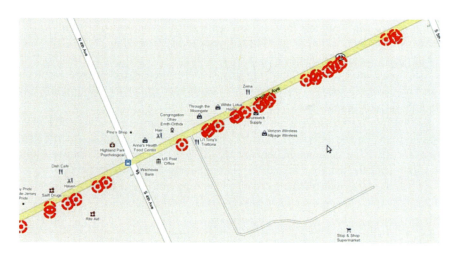

Fig. 8 Map view after offset correction

Fig. 9 Matrix C

$$C_{d,r} = \begin{pmatrix} x_{1,1} & x_{1,2} & \cdots & x_{1,r} \\ x_{2,1} & x_{2,2} & \cdots & x_{2,r} \\ \vdots & \vdots & \ddots & \vdots \\ x_{d,1} & x_{d,2} & \cdots & x_{d,r} \end{pmatrix}$$

Improved System for ParkNet Mobile Network 141

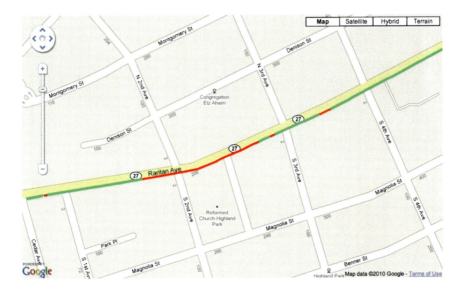

Fig. 10 High-level map view

Presently the ParkNet database contains only the sample data collected during the November 2009 experiment. When sensor-equipped vehicles are deployed live, current data will also be accessible. We have designed two initial views: a high-level, holistic view of parking availability, and a low-level, detailed representation of the locations of valid parking spaces. You can switch views by clicking the "toggle" button.

In the interface, we have two types of view in our interface, one is high-level, and another is low-lever. The figure show above is high level view, when you zoom in the map, it will show some detail information about the road-side parking information, which is parking availability is displayed as a KML file superimposed on Google Maps. Green line segments represent areas that have vacant parking spaces. Red line segments represent areas that are either occupied parking spaces or not parking spaces at all. Such a view may be helpful in planning routes, by enabling a driver to navigate a less occupied street.

Figure 10 is a detailed, point-by-point representation of parking availability for the same location. There are some spot with color along the road. The road view is based on Google map and we use different color to show different condition of the road-side parking position. For example, Green dots represent vacant spots, red dots represent occupied spots (Fig. 11). This type of view may be suitable for drivers in the immediate area, as they seek unique available parking spaces near certain points of interest.

Statistical analysis We conduct statistical analysis of the data collected over the course of 4 weeks during November 2009. To explore how parking

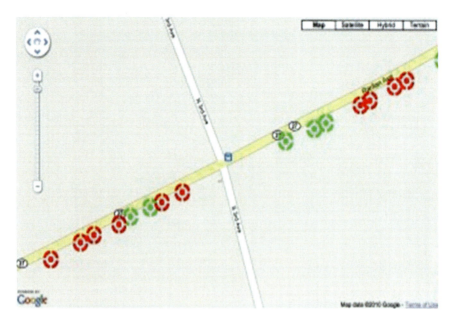

Fig. 11 Low level map view

availability varies as a function of time, we calculated the percentage of spaces available over various time intervals. Data were acquired from sensors while traveling a segment of Route 27, in Highland Park, New Jersey. The segment of road explored contains 57 street-side parking spaces while traveling in the northeast direction. Our analysis is uniquely concerned with data in the northeast direction. The percentage availability was calculated using: Availability $= 1 - (\sum n)/75$. $\sum n$ is the sum of true values as reported by the detection algorithm. For instance, if the detection algorithm identifies 10 parked vehicles on Route 27, $\sum n$ will equal 10. 57 is the number of valid parking spaces. The 23 points on the scatter plot correspond to 23 passes made on the road at various times over the course of 1 month.

Our first analysis measures parking availability over the course of the full month. The x-axis represents the date at which the pass was made, and the y-axis represents the percentage of spaces available during that given pass. The second analysis measures parking availability over the course of 1 week. The same data were used in Figs. 12 and 13, however in Fig. 13, the date at which the data was acquired was ignored, and the x-axis only describes the day of the week during which the pass occurred. "1" means Sunday, "2" means Monday, etc.

The most interesting analysis is Fig. 14, which measures parking availability over the course of 1 day. Again, the same data were used from Figs. 12 and 13,

Improved System for ParkNet Mobile Network

Fig. 12 Parking availability over 1 month

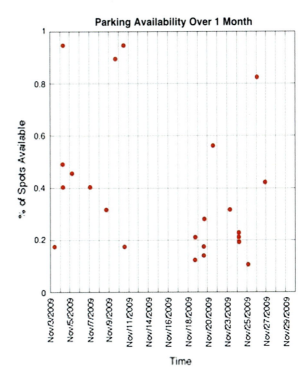

Fig. 13 Parking availability over 1 week

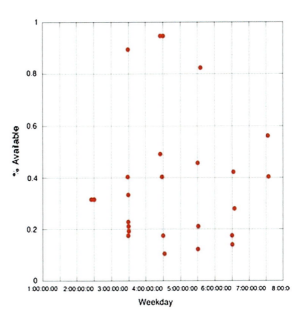

Fig. 14 Parking availability over 1 day

however Fig. 14 is uniquely concerned with the time of day at which the pass was made. Notice that parking availability reaches a minimum at around 11:30 a.m., while availability in the early morning or late afternoon tends to be much greater. Route 27 has many restaurants and businesses, and the times around noon are apparently highly trafficked.

4 Conclusion

We have designed the visualization system for ParkNet, a mobile sensor network for harvesting real-time information about parking availability. First presented in June, 2010, by Mathur et al., ParkNet seeks to ease traffic congestion and frustration by directing drivers towards the known locations of vacant parking spaces. First we present a scalable approach to collecting information about valid parking by populating a database as sensors acquire new information. Next, spot matching can be completed by correcting GPS offset and an application of the Hungarian algorithm. A sample user interface was developed for testing purposes and to illustrate how a possible front-end might appear. Finally statistical analysis of the data reveals that certain trends exist when measuring parking availability as a function of time.

References

1. Mathur J et al (2010) ParkNet: drive-by sensing of road-side parking statistics. In: MobiSys'10. ACM, 2010
2. Mathur K et al (2009) ParkNet: a mobile sensor network for harvesting real-time vehicular parking information. In: MobiHoc'09. ACM, 2009
3. Urban Mobility Report (2007) Texas Transportation Institute, Texas A&M University
4. Smart-parking at rockridge BART station. http://www.path.berkeley.edu/path/research/featured/120804park.html
5. SFPark: about the project. http://sfpark.org/about-the-project/
6. Kuhn HW (1955) The Hungarian method for the assignment problem. Nav Res Logist Q 2:83–97

Application Perspectives for Active Safety System Based on Internet of Vehicles

Ling Chen

Abstract The current mainstream applications of internet of vehicles are to collect vehicle status information, value-added service to customer. And the current active safety systems based on in-vehicle sensors could not give more earlier warnings about potential road dangers, or could just only make an active safety action based on danger detection around vehicles with a short distance. Based on our research for the current mainstream Telematics hardware, various technologies of wireless communication networking, system structures, solutions of V2V/V2R communication, and the integration for the internet of vehicles and automatic active safety system feasibility, This research illuminate the essential conditions for system integration, such as technologies preparation and resources requirements (social resources etc.). And this paper will give the perspectives of its applications base on the marketing analysis data.

Keywords Internet of vehicles (IOV) · Active safety system · V2V · V2R

1 Safety System Application on Various Iov Communication Technique

Mainstream IOV Architecture includes vehicle embedded device/transmission system/data collection and distribution system. Various vehicle embedded information systems apply the below mainstream wireless Tech interfaces

---

F2012-I02-007

L. Chen (✉)
Electrical Department, Pan Asia Technical Automotive Center Co. Ltd, Pudong, China

Mainstream communication technique compare:

Characteristic	WIFI	Public mobile network (2G/3G...)	DSRC
Delay	Seconds	Seconds	50 ms
Range	<100 m	<10 km	1000 m
Mobility	<5 km/h	>60 km/h	<=500 km/h
Rate	11–54 Mbps	<2 M	3–20 M
Band	2.4G/5.2 G	800–900 M/1.9 G–2.1 G	5.85–5.925 G
Standard	IEEE 802.11 b/a/g		IEEE 802.11p

Wireless communication technique and application.

	WIFI	Public mobile network (2G/3G...)	DSRC
Current application	Normal telematics value-added application	Normal telematics value-added application	ETC
Network infrastructure status	1. Small scale Coverage 2. Need setup dedicate access points	Wide coverage	Require specific infrastructure for specific application

1.1 DSRC Application

V2V and V2R Communication via DSRC could transfer the real time data in two directions: When the vehicle close to another one, the coming vehicle can trigger the related collision avoidance system, coming vehicle will transfer the vehicle location/speed etc. simple vehicle driving status information to another vehicle. The two vehicles will warn the potential risk.

When a vehicle safety warning system identify abnormal driving status, it will inform the vehicle around the potential risk and transfer the emergency information vehicle by vehicle in specific model. And abnormal vehicle will communication with road side Unit and transfer related abnormal information and driving status information to data processing center, after data processing and distribution, the data center will provide the more coming vehicles useful traffic data and warning to avoid the potential accident.

Define the dedicated application data interface, vehicle with this system could transfer various vehicle driving status data: location, steering wheel angle, Yaw rate data, speed etc. to vehicles around. And receive the data from the vehicle around, input the data to the vehicle active safety system, the active safety system calculate and warn then driver and take action(speed slow down, brake, turn etc.) automatically.

If the roadside unit is equipped, The vehicle driving status can be transferred to the roadside unit also. The roadside unit will send to the backend information

Application Perspectives for Active Safety System Based on Internet of Vehicles 149

process center after data pretreatment. The backend system will receive the mass vehicle data from the vehicles on road equipped V2R system and follow the specific model to process and analysis the data. The backend system output the traffic status data finally and distribute to the vehicles in order to assist vehicle driving.

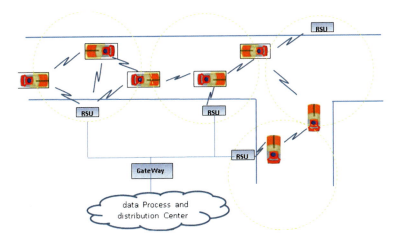

1.2 WIFI Application

Because of the low mobility and long delay characteristic, WIFI is just suitable to wide area information communication which doesn't have high real time requirement. Eg. Distribute the surrounding accident and traffic info via public WIFI hot spot.

1.3 Public Mobile Network Application

Public mobile network is wide coverage and good mobility. Currently many main OEMs use the public mobile network access device (e.g. OnStar, G-BOOK, CARWING, INKANET etc.). Because this kind system equipped 2.5G/3G communication module, it could become the potential data channel for V2X communication. The vehicle could receive real time traffic and traffic control information via this channel. And the vehicle ACTIVE Safety system will warn the driver and take the automatics necessary action base on the data from V2X/V2V and Sensor data.

2 Active Safety Use Case Base on Iov

Use case A:

Vehicle-A communicate with RSU and get accident data after the front turn, the vehicle safety system will drive the brake and powertrain system to reduce speed and avoid action. At same time, the vehicle-A send the warning information related with traffic accident event and driving status(speed slowdown/turn etc.) to the backward coming vehicle-B/C via DSRC. The information will help the coming vehicles avoid accident.

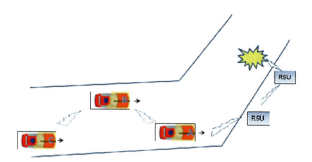

Use cases for V2V and V2R

Use case B:

On the complicated road (e.g. The sharp turn), the vehicle can't see the forward coming vehicle or a driving vehicle to the road cross in high speed from the vertical road.

If both vehicles is equipped with V2V system, the both vehicles can receive the driving status information (location/speed/direction etc.) each other. The vehicle active safety system might warn driver, override steering or apply the brakes automatically.

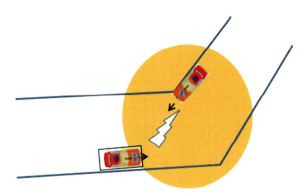

Application Perspectives for Active Safety System Based on Internet of Vehicles 151

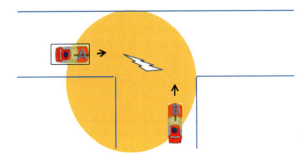

hazard warning
Use case C:
Warn drivers when the vehicle are rapidly approaching a red-light/yellow -light intersection. The vehicle embedded system will communicate with RSU (e.g. intelligent traffic light device) and receive the broadcast traffic light information. The vehicle system will alerts the drivers going too fast and even drive the brake system action for speed slow down.

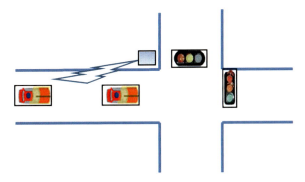

3 Limitation of the System Deployment

So far, due to there is no uniform standards for vehicles data exchange/collection/ distribution and basic infrastructure, therefore this paper couldn't provide a specific analysis for a detail platform building. There also exists safety/tremendous infrastructure investment problem need to be resolved before the system deploy.

Base on IOV, there is a platform to collect and distribute the data ACTIVE SAFETY SYSTEM require. But We lack the necessary standard and infrastructure now. Data usability and data security still have problem. Because of the tremendous cost of infrastructure, U.S. and Europe government encourage the auto industry to do the related system application trail in the small scale trail stage.

Traditional ACTIVE SAFETY SYSTEM vs. ACTIVE SAFETY SYSTEM base on IOV : A traditional ACTIVE SAFETY SYSTEM system is useful no matter if anyone else has it, but ACTIVE SAFETY SYSTEM base on IOV (special for V2V) is useful if many cars around are equipped. It's the network effect system. Creating data and transmission device standards will help cost down and Marketing penetration increase and application network setting up.

Because there are many private vehicle location and driving status information exchange, data security and privacy protection should be considered while technical solution definition. This issue maybe will become one of the barriers for the system deployment e extension. So setting up a security and privacy protection system is main system application development goal. For example: No identifier data transmission would make that the messages from one vehicle could not be linked to particular vehicle. So there's no possibility of tracking patterns or using the data to discern the location of a particular vehicle

4 Conclusion

Vehicle active safety system could enhance its safety execution abilities by integration with internet of vehicles. This integration proposal includes wireless communication solution, standards of transmission and distribution, and related about Infrastructure requirements. This integration system basing on internet of vehicle and active safety system has wide application perspective. The uniform government and related industry plan on the related technique and development will be the necessary support for this application deployment. This integration application will be benefit to the driving safety and intelligent traffic development.

Part III
Telematics, Navigation Systems

Consideration on the Time-Based Evaluation of the Traffic Information Prediction

Ishiguro Yosuke

Abstract In offer of the traffic information to the drivers, it is the big factor that decides quality of the traffic information whether the traffic information and conformity with the real condition experienced by a drivers. The OD time is useful information because the drivers can know the time of arrival to the destination. However only using the real-time traffic information, it is different from the time of arrival to display at origin and the time of arrival at destination because it can't consider the change of traffic condition when they arrive at the destination. So it is generally to raise the benefit of the drivers with the traffic information prediction that can decrease the difference. But now (in particularly the prediction) in evaluating the quality of the traffic information, there are methods to evaluate the traffic jam but not any effective methods to evaluate the traffic information with link travel time. This article suggests the method of evaluating the quality of the prediction with the RMSE in addition to the standard deviation and considers the quality of the traffic information prediction that can offer useful information to the drivers.

Keywords Traffic information · Prediction · Evaluation · Driver's benefit · OD time

F2012-I03-0002

I. Yosuke (✉)
Denso Corporation, Kariya, JAPAN

1 Introduction

By improvement of the telematics technique, many drivers receive offer of the traffic information through a navigator during an driving. The market of the traffic information service changes greatly, and there are many traffic information providers of the private with that of the public, too.

The traffic information that delivers coded data called Traffic Message Channel (TMC) realizes dynamic guidance. The traffic information is offered to the drivers for text messages, or drawing of the traffic jams to a map and gives the drivers route of choice. In addition, the traffic information that received is used in a route search considered the traffic jam and offers to the drivers the fastest route.

Regardless of a type of traffic information offered to the driver, the benefit of the driver is dependent on quality of the traffic information. It is a factor that decides the value of the traffic information service. However, without the characteristic to show such as the location of the control button or the visibility of the screen, drivers cannot evaluate the characteristic of the traffic information service which is based on value in conjunction with the driver's needs by oneself beforehand. Therefore it is possible only to depend on the reputation of the last supplier (for example, auto manufacturer) of the traffic information.

Furthermore, the last supplier (auto manufacturer) of the traffic information faces the problem that they evaluate quality of their traffic information service objectively to maintain reputation.

And when they can choose several traffic information providers and their original service, they have a problem of the cost.

Therefore, the objective high quality evaluation becomes important almost as same as the evaluation of the conventional automobile function (for example, fuel consumption and engine torque). In addition, the drivers receive the "free" traffic information service (of the public). The "pay" traffic information service that traffic information providers (of the private) offer to the drivers must offer the additional benefit to the drivers.

The present objective quality indices relate most the delivery of traffic information and establishes about of the quantity or the technical characteristic. For example, it is minimum indices that relate the time and space such as the powerfulness of the receiving sensitivity and delivery velocity, a delivery area to offer the traffic information to the drivers.

However, the quality indices don't pay its attention whether the traffic information and conformity with the real condition experienced by a drivers.

However, unsureness of the content is a factor that decides both use satisfaction and whether the drivers receive the traffic information service.

This article gives examples of the method that decide the quality indices of the traffic information to solve upper problems. And this article considers about the objects of examples of the method and suggests efficiently method that decides the quality indices of traffic information to solve the objects. The method pays its attention to travel time per link of traffic information prediction offered to the drivers.

2 Generating Traffic Information

Generally the traffic information such as the text message, drawing of the traffic jam to a navigation map and fastest route is generated in "traffic information service system". That consists of three functions that are (1) the instrumentation of the traffic condition (2) disposal of instrumentation data (generation of the real-time traffic information and generation of the traffic information prediction) (3) the offer of the traffic information.

2.1 The Instrumentation of the Traffic Condition

There is a method to collect drive phase from a probe vehicle equipped with various sensors such as a vehicle speed sensor or the GPS and to collect the data by always measuring the traffic condition by the sensor that set at road side to traffic information center.

Because probe vehicles can collect traffic conditions in the road without a sensor, the instrumentation of the traffic condition by the probe vehicles can expect an enhancement of the cover area of the traffic information. In addition, because a probe vehicle that only setting various sensors on a vehicle can reduce the instrumentation cost of the traffic condition in comparison with a road side sensor, many traffic information provider of the private collect traffic conditions by probe vehicle. However, because there is the problem not to be able to always measure the traffic condition like a road side sensor, the instrumentation data collected by probe vehicles affects quality greatly.

2.2 Disposal of Instrumentation Data

In the generation of the traffic information (prediction), the traffic information center processes instrumentation data into the traffic information available with a navigation by gathers and processes statistic data that measured by road side sensors or probe vehicles. Particularly in generation of the traffic information prediction, there are large differences in the quality of the traffic information by because a characteristic of the traffic information prediction changes by a method of the processing of the instrumentation data that gathered by them.

2.3 The Offer of the Traffic Information

The offer of the traffic information offers the traffic information to the terminals such as navigations and portable phones, web processed by traffic information center.

In addition, the traffic information is offered in a traffic jam per link or link travel time. The traffic jam is classified depending on a velocity level of traffic information processed in the link. For example, that are classifications that if velocity is less than 30 km/h, it is "stationary traffic", if velocity is less than 50 km/h, it is "traffic jam", and more then, it is "free flow" in highway. It is used for the drawing of the traffic jam to the navigation map or the text message.

Link travel time that processes statistically the instrumentation data is used mainly by the representation in the time of arrival to the destination or searching fastest route.

Because the traffic condition always changes, only using the real-time traffic information, it is different from the time of arrival to display at origin and the time of arrival at destination. That reason is that it can't consider the change of traffic condition when they arrive at the destination.

So several providers predict traffic condition by gathered the instrumentation data, and they raise the benefit of the drivers with the traffic information prediction that can decrease the difference.

3 Existed Quality Indices for Traffic Information and Objectives

In the traffic information argued before, It influences the benefit of the drivers greatly whether the traffic information and conformity with the real condition experienced by a drivers. There is QKZ [1] that is the quality indices of the traffic information.

QKZ is the quality indices of traffic jam that offered to the drivers and QKZ has two indices.

Quality index one (QKZ1), the detection rate, describes the degree to which the traffic information coincides with the actual traffic condition and is calculated from the ratio of "stationary traffic" of the area of the intersection (between the traffic information and the actual traffic condition) to the area of the actual traffic condition.

Quality index two (QKZ2), the false alarm rate, describes the proportion of the traffic information that is not relevant to the "stationary traffic", i.e. the proportion of the area of the information that lies outside the "stationary traffic" area.

You make the two-dimensional description of two quality indices and can interpret the quality of the traffic information.

However, QKZ method can measure quality of the discrete traffic jam degree, but cannot measure quality for continuous link travel time.

In an offer for the link travel time (particular in the representation of the time of arrival of the destination), the traffic information prediction can raise the benefit of the drivers. So the quality index for the prediction of the link travel time is important. The quality index by Root Mean Square Error (RMSE) is common for a method to measure quality of the predictive link travel time, but there is the case that result changes by the choice of the evaluated sample in RMSE.

That reason is that if the fluctuation range of the traffic condition is different in evaluated link or time, the size of error in prediction change. For example, in a certain link, as for the error in prediction of the link travel time of 3:00 a.m. and the error in prediction of the link travel time at 8:00 a.m., error in prediction at 3:00 a.m. shrink. Because generally the middle of the night tends to have free road and the traffic condition is less changing, the traffic condition is easy to predict, the error in prediction shrink.

In other words, it may be said that the quality of the traffic information prediction is good if you increase evaluated samples of the prediction of the link travel time at 3:00 a.m.

To prevent above, you can prepare evaluated samples in all time. But in the scene of the representation of the time of arrival, when the change of the traffic condition is large and then the error in the prediction of the link travel time is small, the drivers can receive larger benefit. So it has good quality of the prediction that error in prediction of the link travel time is small when the change of the traffic condition is large.

And the evaluation adds weight to predicted error. Based on the QKZ and the driver's sense, when the change of the traffic condition is large and the traffic condition is "stationary traffic" and the passage frequency of the drivers is high, the weight raises.

Because when the traffic information offers to the drivers in that situation, the drivers can receive large benefit.

On the basis of the above, the following chapter argues about the quality of the prediction of the link travel time with numerical formulas.

4 Quality Indices for Traffic Information Prediction

RMSE defined the following (1) that expresses predicted error for an index of the quality of the traffic information prediction. Then the size of predicted error is expressed the link travel time per length of the link, not only link travel time. The reason is because it can compare the size in the same index by normalization the error in order that the size of the predicted error is different by length of links.

$$RMSE_{S,T} = \sqrt{\frac{1}{n} \cdot \sum_s \sum_t \left(\frac{ltt_{s,t}}{l_s} - \frac{\hat{ltt}_{s,t}}{l_s}\right)^2} \quad (1)$$

$$\mu_{S,T} = \frac{1}{n} \cdot \sum_s \sum_t \frac{ltt_{s,t}}{l_s} \quad (2)$$

$ltt_{s,t}$: real-time link travel time of link "s" and time "t" (3)

$\hat{ltt}_{s,t}$: predicted link travel time of link "s" and time "t" (4)

l_s : length of link s (5)

$s \in S$: The unit of spatial division (ex. links)
n_s : The number of samples (6)

$t \in T$: The unit of time division
n_t : The number of samples (7)

$$n = n_s \cdot n_t \quad (8)$$

For another index to evaluate the quality of the prediction, the standard deviation (σ) is defined as follows (9).

$$\sigma_{S,T} = \sqrt{\frac{1}{n-1} \cdot \sum_s \sum_t \left(\frac{ltt_{s,t}}{l_s} - \mu_{s,t}\right)^2} \quad (9)$$

The RMSE calculates it with the units that divided adequately spatially and time [shown in (6) and (7)]. For example, by calculating RMSE with a unit for five minutes and for a link, it can recognize visually the quality of the prediction (Cf. Fig. 1). And you can know in which link or time the quality of prediction is high.

Making the units smaller, you can know the quality of the prediction more detail. However, there is the problem that the reliability of RMSE and the standard deviation to mention later falls down because the number of the samples decreases. It is entrusted how small you divide the units.

Consideration on the Time-Based Evaluation   161

Fig. 1 Two dimensional diagram (RMSE and standard deviation)

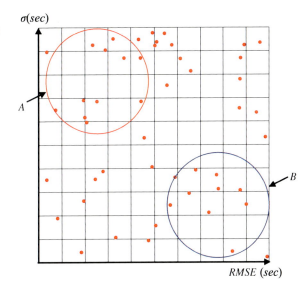

In order to interpret the quality of prediction, two-dimensional diagram is shown (Fig. 1). The two indices are the RMSE and the standard deviation.

At first it is clear that the quality of the prediction is good if the RMSE is small. In addition, the quality of the prediction at the part to show in area A (Fig. 1) can be superior to the case that the standard deviation is small. This is based on that it can give the drivers big benefit that the error of the predicted link travel time is small when the change of traffic condition is large. In other words, it is the "bad" quality of the prediction at the area B that the error in prediction are large though there are few change of the traffic condition.

The comparison between the qualities of the different predictions is enabled by the representative value that calculated the RMSE with weight.

That is the representative error of the prediction that has the same input with weight that considered the value of the traffic information prediction and the sense of the drivers.

The weight that considered the value of the traffic information prediction is adopted so that the error of the prediction grows large when the standard deviation is large. In addition, when the drivers are easy to receive the traffic information, the weight that the sense of the drivers is adopted so that the error of the prediction grows large. For example, in the time that the length of the trip distance is long, or at the link that has mach traffic volume, the weight grows large. So you can evaluate the quality of the prediction considered the utility value of the traffic information of the drivers.

By using that weight, the calculation of the representative value of error in the prediction that gained weight based on value of the traffic information prediction and the sense of the drivers is enabled. Then you can compare the qualities of the different predictions.

And by using this method, interpretation of the quality is enabled about the prediction any time after.

5 Summary and Outlook

This article suggests the method of evaluating the quality of the prediction with the RMSE in addition to the standard deviation. The method is based on that it can give the drivers big benefit that the error of the prediction is small when the change of traffic condition is large. And by using the calculation of the representative value of error in the prediction that gained weight based on value of the traffic information prediction and the sense of the drivers, you can compare the qualities of the different predictions.

In the future, it is necessary that we have to evaluate based on true data, and inspection of the fairness of financial statements of the method of evaluation. In addition, in the evaluating of error in the prediction, it is necessary to make the error in the prediction which considered correlation with the adjacent link.

Reference

1. Bogenberger K (2003) The quality of traffic information. Airsage. http://www.airsage.com

Barge-in Implementation Method for Multi-CPU In-Vehicle Speech Recognition System

Naoyori Tanzawa and Yoshikazu Inagaki

Abstract The objective is to implement barge-in function for multi-CPU in-vehicle speech recognition system. The barge-in can allow user utterances during the guidance prompt from the system. Barge-in function requires two input audio data. The one is speech signal from microphone. The other is the original guidance prompt data. There are two major problems to implement barge-in. The one is sampling frequency synchronization, because the frequencies are usually different between two input audio data. The other is input timing synchronization, because the timing gap between two input audio data must be within plus or minus 7 ms based on the specification. We have adopted digital hardware converter for sampling frequency conversion. Hardware processing makes much smaller signal input delay than software processing. In addition, digital processing keeps the performance of barge-in compared to analog processing, because it does not cause the quality deterioration. We store the reference data right before audio data input for barge-in module, because the variabilities by the reference data transmission between two CPUs can be removed. Regarding the microphone input data, we reduce the variabilities by synchronizing the microphone input request with the guidance prompt play request. The guidance prompt taken by microphone is passed into barge-in module with the reference data one by one in the smallest unit sequentially to keep real-time processing. The implementation method was validated in terms of the design. The presented design has been implemented and evaluated. Conclusively, it is validated based on the evaluation in terms of the input timing gap between two input audio data into barge-in module and the performance of barge-in function. This study was specified for multi-CPU in-vehicle speech recognition system. The prerequisite is that the full reference data

F2012-I03-004

N. Tanzawa (✉) · Y. Inagaki
Denso Corporation, Kariya, Japan

must be on memory to keep real-time processing. Our method is able to solve the two major problems by hardware sampling frequency conversion and the proposed input timing synchronization method. In addition, this method keeps the real-time processing of speech recognition even if we added the barge-in function as the preprocessing of speech recognition. We proposed a barge-in function implementation method for multi-CPU in-vehicle speech recognition system. It solves problems regarding sampling frequency and input timing synchronization of two input audio data to adopt barge-in function. This method does not make any latency performance issues of speech recognition.

Keywords Speech recognition · Barge-in · Multi-CPU · Sampling frequency synchronization · Input timing synchronization

1 Introduction

The improvement of recognition rate is an endless challenge in the speech recognition field. The task achievement rate using speech recognition is 67 % according to research. 21 % of error causes are from interrupting speech. This is the task the barge-in function tries to solve. Figure 1 shows the ideal situation and the real situation in terms of the beginning of the speech. An interrupting speech happens in the (b).

The barge-in mechanism is shown in Fig. 2. There is echo cancellation technology for telephone system. It helps to avoid the reflection of own voice through phone. The barge-in function makes it possible to remove guidance prompt in the audio data through the microphone by applying the echo cancellation technology. As a result, only user speech data is input into speech recognition engine, although the user makes a speech during guidance prompt.

The objective of this method is to implement barge-in function for multi-CPU in-vehicle speech recognition system in accordance with the restricted environments. The system architecture we adopted for barge-in is shown in Fig. 3. It consists of two CPUs. Speech play module is on CPU-1. Speech recognition and barge-in modules are on CPU-2. The amplifier is independent from the head unit, and it is connected with MOST communication. Microphone input data is transmitted as shown in the figure.

Barge-in function needs two input audio data. The one is speech signal from microphone. It is mixed with guidance prompt from vehicle speakers. The other is original guidance prompt data as reference. There are two major problems to adopt barge-in for multi-CPU in-vehicle speech recognition system. The one is sampling frequency synchronization, because the frequencies are different between two input audio data. The other is input timing synchronization, because the timing gap between two input audio data must be within plus or minus 7 ms based on the barge-in module specification.

Barge-in Implementation Method

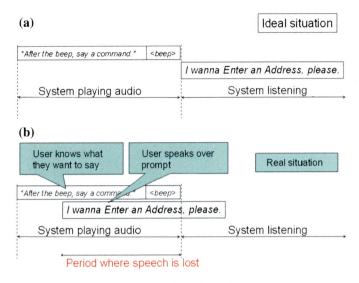

Fig. 1 Interrupting speech. **a** Ideal situation, **b** Real situation

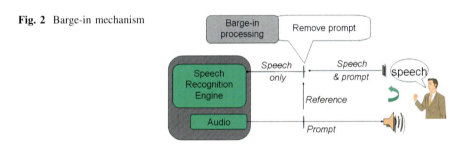

Fig. 2 Barge-in mechanism

2 Methodology

2.1 Sampling Frequency Synchronization

The sampling frequency of the original audio data is not necessarily the same as the sampling frequency of the microphone input data. The sampling frequency for microphone data is 16 kHz in our environment. This is defined by speech recognition engine specification. On the other hand, the rate for reference data is 18.9 kHz. This is determined based on the balance between the quality and the data size. Accordingly, we have to change the rate of reference data from 18.9 to 16 kHz. We considered three ideas shown in the Table 1.

These are the combination of the transmission and the conversion methods for reference data. The advantage of digital transmission is that it has no degraded audio quality. On the other hand, the advantage of analog transmission encourages the improvement of the barge-in performance, because the same ADC processing

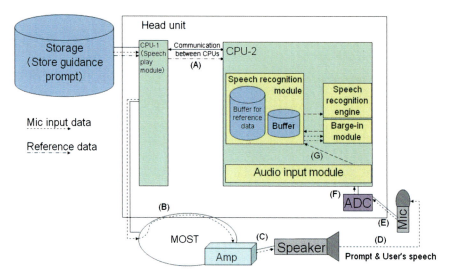

Fig. 3 System architecture

Table 1 Three ideas for sampling frequency conversion

Ideas		Advantage	Disadvantage
1	Transmission: digital	Keep audio quality	Processing time
	Conversion: software	No hardware added	Variability
2	Transmission: digital	Keep audio quality	Need additional software
	Conversion: hardware	Quick processing	
3	Transmission: digital	Performance	Converters are needed
	Conversion: hardware	Quick processing	

is assumed with microphone input data. However, the clock source has to be the same between DAC and ADC to avoid the mismatch of acoustic wave. Regarding the conversion, hardware requires additional cost. However, it avoids processing time variabilities. As a result, we decided to adopt the second one, because it is feasible and the impact for timing synchronization is small.

Figure 4 shows the routing for the sampling frequency conversion in CPU-1. The same path between reference data and microphone input data is kept to save the processing load and avoid the mismatch of acoustic wave.

2.2 Input Timing Synchronization

We have to input two audio data within plus or minus 7 ms for barge-in module. It is based on the barge-in module specification. Firstly, we considered that the speech recognition module synchronizes the input timing when it requests the

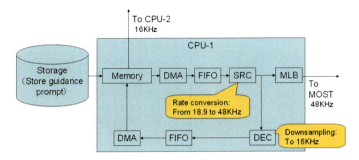

Fig. 4 Routing for sampling frequency conversion

guidance prompt play. The advantage is that it does not require any big change in terms of software. The disadvantage is that it causes variabilities on all paths to the barge-in module. On the other hand, we considered the way to detect the beginning of the speech by speech recognition engine. However, the accuracy does not satisfy the requirement. Accordingly, we considered synchronizing the input timing by the speech recognition module. The request and audio data are passed as follows. Figure 3 also shows the paths.

Speech recognition module = (A) => Speech play module = (B) => Amplifier = (C) => Speaker = (D) => Microphone = (E) => ADC = (F) => Audio input module = (G) => Speech recognition module

The processing time of the path B, C, D, E, F is stable. Accordingly, the processing time of the remaining paths should be verified in the target environment.

We have considered that reference data is stored in the buffer in speech recognition module as Fig. 5 shows. The microphone input data is stored with each unit. It is passed into barge-in module with the same part of reference data. This idea can solve the variabilities happened on each path of reference data. It avoids the impact for the latency of speech recognition, because all the reference data is buffered in advance. Accordingly, we adopted the idea.

3 Results and Discussions

The implementation method was validated in terms of the design. The presented design has been implemented and evaluated. Conclusively, it is validated based on the evaluation in terms of the input timing gap between two input audio data into barge-in module and the performance of barge-in function. This study was specified for multi-CPU in-vehicle speech recognition system. The prerequisite is that the full reference data must be on memory to keep real-time processing.

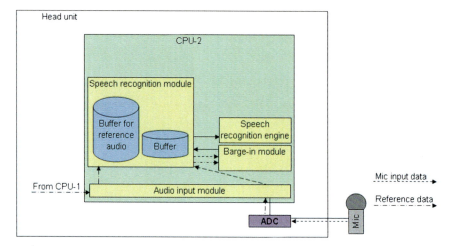

Fig. 5 Buffering for reference data

4 Conclusions

We proposed a barge-in function implementation method for multi-CPU in-vehicle speech recognition system in this paper. We have two problems to implement barge-in function such as sampling frequency and input timing synchronization. We solved first problem using digital conversion by hardware processing. Regarding the second problem, we considered that the speech recognition module synchronizes the timing when it requests the guidance prompt play. Reference data is stored in the buffer in speech recognition module to avoid the variabilities happened on each path of reference data. We will verify the variability is within plus or minus 7 ms implementing the design.

Seamless Traffic Information in Dynamic Navigation Up to Date

Regina Glas, Heidrun Belzner, Tim Lange, Irina Koller-Matschke and Richard Wisbrun

Abstract Preciseness and reliability of traffic data as an input for GPS navigation units systems means a competition relevant advantage for system provider. It is shown that many data sources for those systems are potentially available today. Most of these sources however either lack in penetration or do not provide sufficient network coverage. In conclusion this means that relying solely on a single data source will not be sufficient to build a high quality navigation solution. This paper gives an overview of potential data sources and elaborates their pros and cons. Relying on existing infrastructure and vehicle probe data "premium quality" can be achieved, on highways at least. This does not fulfil the customer's expectance of a premium navigation system if arterials are not covered in the same quality as on highways. Significant reduction in travel time is the only valid criteria of quality of recommended changes in routing. Additionally real time traffic information and dynamic route guidance presumably leverages the highest potential especially on secondary road networks within large metropolitan areas. In order to continuously fulfil customer expectations BMW Group engineers collaborate with service providers and engage in various research projects to bring together multiple data sources and develop new data collection methods.

Keywords GPS navigation systems · Routing · Stationary sensors · Mobile Sensors · Traffic management

F2012-I03-005

R. Glas (✉) · H. Belzner · T. Lange · I. Koller-Matschke · R. Wisbrun
BMW AG, Munich, Germany

1 Motivation

With constantly rising traffic volumes and surging numbers of GPS navigation units requirements for quality of the displayed traffic information increases. Preciseness and reliability of traffic data as an input for those systems means a competition relevant advantage for GPS navigation system provider (see Fig. 1). Today still large segments of the road network are lacking in sufficient coverage and correct data in correct travel time prediction. Frequently traffic incidents are inaccurately localized or too late detected. In order to guarantee high product quality, BMW Group engineers have been seeking new methods of gaining access to seamless and accurate traffic data.

2 Stationary Sensors

There are a lot of stationary detectors that differ in technology and the possibilities of evaluation.

2.1 Traffic Flow Measurement

Stationary sensors are often inductive loops, permanently installed in the pavement. But there are also optical, acoustical, and micro wave systems. The systems measure the number of vehicles on a section and their speed. A typical aggregation interval is 1 min. Normally it is distinguished by cars and trucks. Parameters like traffic volume and travel times can be measured indirectly on base of estimated by traffic models [1].

The data are used prevailing as basis for traffic monitoring and traffic management besides the generation of traffic information. With a good coverage of sensors, e.g. in Germany generally every 1–3 km on motorways, a quality of 80 % detection rate of traffic congestion is assumed to be possible. One main problem is that the data are often inaccurate or completely be absent, because there of hazard spots like road works or out of order due to the construction works in the majority of cases. Further problems are high maintenance and repair costs as well as weather sensitivity.

Besides the strategic inductive loops on highways and motorways there are traffic light detectors in urban areas. Those are mainly installed at intersections to control the traffic signal phases dynamically. Those traffic sensors are mostly simply only traffic flow. Other measurements are not commonly of use, because of cars holding on the detectors at the intersection in front of a traffic signal. Therefore the data is barely to be used for generation of traffic information.

The high distribution of those static sensors opens up chances to close the often existing information gap in urban areas.

Unfortunately that information was until nowadays not planned to make publically accessible. The data are owned by the local authority, which is not in

Fig. 1 Congestion concerned information needs of travellers in cars

duty to publish any control information openly, and even those data are not always sent to a central server, because many of the detectors are only connected to the local control system at the intersection. Therefore there is a bad to grip for traffic information.

2.2 Travel Time Measurement

Automatic number plate recognition (ANPR) detects travel times between sections. The license number is collected encrypted and analyzed anonymously. ANPR could also be used for toll collection.

The experience of the ANPR techniques is positive. Due to infrared cameras and reflecting license numbers the detection rate is above 90 %. Thus, the technical constraints are given to generate a robust, comprehensive, and long period data base. In Munich the BMW Group has installed an ANPR system together with further partners in the research project wiki.[1] Every minute exact traffic state data on the main road network are provided. Because of the high investment costs to install the system, ANPR is adopted only in a few networks and areas until now.

---

[1] Wiki—Wirkungen von individueller und kollektiver ontrip Verkehrsbeeinflussung auf den Verkehr in Ballungsräumen.

Fig. 2 Shows an overview on stationary and mobile sensor systems

In countries with toll collection the system is sometimes also used to get travel times and traffic information. This is done by analyzing the entering and exit times based on ticketing times, number plate recognition or RFID (e. g. in Italy, Dubai). Other systems are based on the Bluetooth signal of mobile phones insides the cars [2].

Beacon systems use two-way communication between the car and the infrastructure. Cars send travel times to the beacons and get back traffic information generated in the traffic management center. Those systems are infrastructure-based and only equipped cars participate on the data collection and communication and infrastructure costs are present. Today beacon systems are mainly in place for electronic toll collection or for location of public transport vehicles like busses. Countries with a high risk of earthquakes (e. g. Japan) use that kind of systems because it has maintenance benefits compared to techniques like inductive loops that are fixed in the pavement.

3 Mobile Sensors

Mobile sensors usually don't have high maintenance costs compared to the effort of static detectors (see Fig. 2). Additional sections based traffic data can be collected by using vehicles. Due to the manner of collecting data these sensors are described as Floating Cars (FC) or Probes. The rapid evolution in the area of cellular mobile communication provided the basis for a commercial use of cellular phones for traffic information. This is called Cellular Probe Data.

3.1 FCD: Floating Car Data

Using cellular mobile communication vehicles are able to detect and send data via the entire mobile network. That is nearly everywhere and every time because of the business needs of telecommunication providers. Using their infrastructure a network-wide traffic state detection is made possible. This induced that the number of vehicles, normally via the high coverage of mobile phone or radio communication carriers in cars, that collect data highly could be increased. In Germany traffic data collection is strongly based on stationary data. Whereas in countries as the US or Great Britain fleets or taxis are used.

Fleets and taxis have the advantage that they have high mileages and, in contrast to private cars, no aspects of data privacy have to be respected. Also these cars already possess communication hardware for sending their position to a centre. But these means of transport show some characteristics that have to be considered by analyzing the motion for traffic state detection. E.g. there exist certain speed limit restrictions for fleets and the motion of taxis that do not always correspond to those of private vehicles. This requires plausibilization.

Since 1999, the BMW Group deployed vehicles to offer specific services with integrated mobile communication infrastructure, which also are used as sensors for improving the traffic information service. Each floating car is equipped with a GPS receiver to enable to offer location based services inside the specific car and anonymously to track section speed estimates at the car positions for traffic flow information. Vehicle data is transmitted anonymously to a traffic data centre by a mobile telephone module.

A later version of that FCD algorithms compare their travel times onboard with expected travel times. Currently they only send messages when exceeding certain congestion thresholds, because the message frequency and volume has to be limited.

The so called GATS[2]-Standard which is the basis for the communication to the centre was developed in the mid 1990s by Mannesmann Autocom and Tegaron Telematics. Today there are about 40,000 FCD vehicles in Germany sending their data mainly on motorways.

3.2 XFCD: Extended Floating Car Data

The increasing number of electronic components enables the use of additional data of the vehicle bus systems (see Fig. 3).

Besides data originated from indicator or accelerations, also information from other vehicle sensors or driver assistance systems can be considered. Therefore, the BMW Group has refined the FCD approach. On the in-vehicular telematic

---

[2] Global Automotive Telematics Standard.

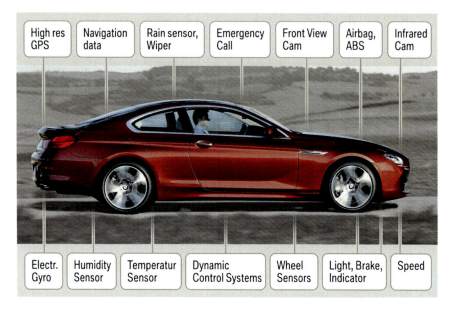

Fig. 3 The intelligent XFCD software determines the different aspects of the actual traffic state and integrates single sensors to a distinct signal of a higher value

platform the intelligent XFCD software determines the different aspects of the actual traffic state. With XFCD more information about the traffic state can be generated and this even more precisely than with FCD. Also interpretation logic is already filtering and fusing data in real-time inside the vehicle. To further reduce communication costs only fused messag.es are only sent in case of specific events that are relevant for the other costumers.

3.3 Cellular Probe Data

In industrialized countries there are cellular phones in most of the vehicles. They have the advantage of a high penetration rate. Additionally, no common equipment investment respectively only minor investments are necessary to build up an infrastructure coverage, because the data providers have other reasons to acquire a device which is in this case also a sensor. Depending on the access to the data at the mobile network operator data from on-call or off-call cellular phones can be used. Although having a high penetration rate the quality of localization is minor compared to FCD, also (but to a more limited measure) with XFCD. Particularly in urban areas it is difficult to distinguish the motion of cellular probe data (not with XFCD) in vehicles from those on bicycles, people leaving a car or in public transport. Due to the building density the GPS detection has an important uncertainty. Before adoption in urban areas these difficulties have to be solved by using high-performance data processing procedures. However, in non-urbanized areas cellular probes show a high potential.

4 Traffic Management

The described stationary and mobile data sources are common that automatic detection systems are used to get traffic flow, traffic density, traffic state, or travel time information. In the following information sources of different nature are described. Guidelines and store of knowledge of the public authorities and road administration are in the focus.

4.1 Traffic Management Strategies

Traffic management strategies are elaborated routing concepts of public authorities. Based on the concepts defined alternative routes are assigned manually or automatically to the primary route in case of traffic congestion or events. Mostly the detection of the congestion is done by stationary sensors, the events were normally planned. The resulting recommendation of the alternative route is communicated predominantly by collective information like dynamic traffic signs to the drivers. Today in navigation systems this information is not used, because of the lack of integration. The navigation route is calculated autonomously, maybe considering only actual traffic information. Thus, contradictory and therefore confusing and even counterproductive information could result.

A new approach is done by the strategy compliant onboard routing. An adjustment of individual onboard navigation and traffic management strategies is implemented. The result is a win/win situation. The individual driver profits on the more globally strategy compliant route because the navigation considers local anomalies (e. g. modified traffic routing due to events) that are not provided in the general traffic information about local hindrances. Furthermore, optimized traffic control systems (e. g. green waves, extra lanes, etc.) are often activated on the defined alternative routes to enhance the capacity and to make a free and faster driving possible. The general public profits on the strategy compliant route because the driver take defined alternative routes and they do not drive on roads that are not build for high traffic volumes.

In cooperative research projects such as INVENT[3] and DMOTION[4] the BMW Group has realized a strategy compliant routing in a prototype (see Fig. 4).

---

[3] Intelligenter Verkehr und nutzergerechte Technik—Intelligent traffic and user-friendly technology.
[4] Düsseldorf in Motion.

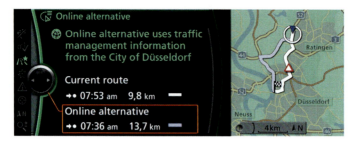

Fig. 4 Suitable vehicle sensors for an intelligent traffic state detection with XFCD

Thereby the strategic routes—defined by road and city authorities—were transmitted in real-time to the vehicle and integrated into the onboard navigation.

The data availability is yet limited to a small number of collaborating major cities and regions, which have already elaborated strategic routing concepts, which they want to be wider communicated. Transferring the available data in a standardized and affordable way to service providers is a challenge. The benefit for the drives depends strongly on the quality of the routing concepts and of the corresponding traffic detection on the main routes, as well as on the optimized traffic control measures on the alternative routes.

4.2 Road Work and Event Information

For some years the BMW Group already integrates long term road works into its navigation systems via RDS-TMC.[5] Those infos are based on the German BIS[6]—a national system which stores all road works on highways that last longer than 8 days. Thereby drivers are warned of hazards and bottlenecks. The navigation system calculates a faster route, in case sensible alternative route options are available. Because the BIS only delivers info about long-term road work on highways, blind spots remain concerning short-term or moving road works and or those on minor roads.

The BMW Group elaborated new approaches for short-term road work within the research project DIWA[7] by installing GPS and GSM modules into warning trailers. When a warning trailer was activated on the road, a warning message was automatically created and sent to the service provider who forwarded this info to drivers via TPEG.[8] The navigation system then warned the driver on-time and with a higher precise about the location (see Fig. 5).

---

[5] Radio Data System—Traffic Message Channel.

[6] Baustelleninformationssystem des Bundes und der Länder—Road work information system.

[7] Direkte Information und Warnung für Autofahrer—Direct info and warning for drivers.

[8] Transport Protocol Expert Group—next generation of RDS-TMC.

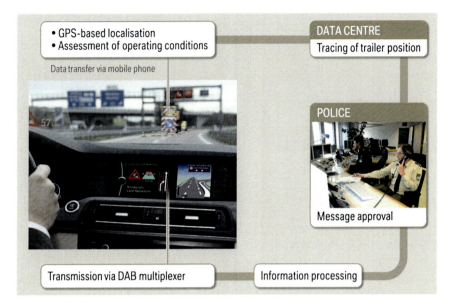

Fig. 5 Stationary and mobile detector systems, area of detection and data quality (size of symbol indicates significance of source)

The blind spot on minor roads in urban areas can be filled by acquiring data on long-term road works from traffic management centers that are becoming common in larger cities. Besides this data information on traffic relevant events—such as sporting games and festivals—is useful to avoid a resulting disturbance in a wide area. This event data is most often also available in traffic management centers or other administration departments precise in time and position.

However, heterogenic and proprietary systems on the side of public authorities make it difficult to transfer their knowledge in a standardized and cost effective way to service providers.

5 Conclusion

This article shows that many data sources are potentially available today. Most of these however either lack in penetration or do not provide sufficient network coverage (see Fig. 6).

In conclusion this means that relying solely on a single data source will not be sufficient to build a high quality navigation solution.

It is fairly realistic to assume that by relying on existing infrastructure and vehicle probe data "premium quality" can be achieved, on highways at least. This however will not fulfill the customer's expectance of a premium navigation system if arterials are not covered in the same quality as on highways. Suggested detours

data source	present availability	quality
traffic flow measurement	• mainly highways, often in context of traffic control systems • partly major inner-city roads	• exact measurement of local speeds • interpolated segment-based speeds • dependent on detector density
travel time measurement	• mainly highways • pilot urban installation in Munich	• very good
FCD	• all road types • ~ 200.000 vehicles in Germany	• exact measurement of local speeds • fuzzy • dependent on penetration
XFCD	• all road types • BMW rollout pending	• very good
cellular probe data	• highways • major roads in rural areas • ~ 80 mio. mobile phones in Germany	• high fuzziness and high rejection rate • dependent on On-Call and Off-Call • data processing very complex
strategic routes	• metropolitan areas • highways	• dependent on local concept and realization
road work data	• in Germany: only highways	• in Germany: good
event data	• metropolitan areas	• dependent on data actuality

Fig. 6 Table with overall view on availability and quality on data

due to traffic congestion on the original route especially on arterials are only of benefit if a significant reduction in travel time is achieved. Additionally real-time traffic information and dynamic route guidance presumably leverages the highest potential especially on secondary road networks within large metropolitan areas. In order to continuously fulfill customer expectations BMW Group engineers collaborate with service providers and engage in various research projects to bring together multiple data sources and develop new data collection methods.

References

1. Belzner H, Lange T, Koller-Matschke I (2010) Dynamic navigation: seamless traffic information already today?" ATZ—Automobiltechnische Zeitschrift: 2010-08
2. Young S (2008) Bluetooth traffic monitoring system, University of Maryland Bluetooth Brochure Rev3. http://www.catt.umd.edu/documents/UMD-BT-Brochure_REV3.pdf

Traffic Information and Individual Driver Behaviour

Irina Koller-Matschke, Heidrun Belzner and Regina Glas

Abstract *Research Questions* In order to exploit the capacity of the existing infrastructures, many investigations for traffic management systems have already been conducted. Despite of high investigations the impact of different management systems is only known partially. Thus, the following questions are addressed in this study: How should traffic information be designed to correspond best to the individual sensation? Is there an appreciable impact on route choice behaviour by radio traffic service, variable traffic signs, or navigation systems? What are the options for operators to reduce traffic flow on overloaded streets without traffic diverts on other routes? *Methodology and results* The mentioned questions were addressed in the research project "wiki—impact of individual and collective ontrip traffic management on the traffic flow in urban networks". The project was focused on individual and collective traffic management systems in urban networks. Therefore three different types of systems were taken into consideration that affects route choice: radio traffic service, variable traffic signs, and navigation systems. To quantify the impact, the traffic behaviour of car drivers was evaluated by different observation methods in a field test in Munich: ANPR—automatic number plate recognition, GPS-logging and interviews, driving simulator. The test field covered the northern part of Munich, including highways as well as urban streets. Overall 51 ANPR cameras had been installed on 22 control sections. But the project was not only focused on the objective traffic state but also on individual factors for route choice behaviour. The comprehensive amount of 300 people had been tracked for 8 weeks by GPS-logs and interviewed about their individual aspects and motivations. The test revealed that the route choice behaviour was mainly influenced by the travel time. The test persons showed high preference to

F2012-I03-007

I. Koller-Matschke · H. Belzner · R. Glas (✉)
BMW AG, Munich, Germany

stay on the main route and were only willing to change to the alternative route if a considerable threshold of travel time increase was obtained. Thereby, the individual behaviour was mainly influenced by radio traffic service and the traffic state displayed in the navigation map. Particularly another test in the driving simulator with additional 40 test persons gave knowledge of subjective motivation and also regarding the influence of familiarity with the network. The test persons rated the traffic information and classified colour scales during and after the simulator drives. This was the base for significant research results of a user friendly layout of traffic information. *New aspects and limitations of the study* Route choice is affected by many aspects. Especially the individual behaviour has high impact but is sophisticated to study. Thus, this study was focused not on theoretical aspects or simulation but on a widespread empirical field test. The combination of different observation methods made it possible to collect the determining factors on route choice and to give indications for operators how to provide traffic information best. The test field covered highways as well as urban streets. Thus, the research results gave new indications how to design traffic information especially in urban areas and in networks that the drivers are familiar with. *Conclusion* The study gives significant results what kind of individual aspects the route choice behaviour affects based on a wide field test. On this basis recommendations are made which traffic management systems have the best influence on the driver behaviour. In addition it was analysed how traffic information should be designed.

Keywords Traffic information · Navigation · Route choice · Traffic state · Traffic management system

To analyse route choice behaviour a lot of information on traffic is needed. For example origin and destination of a trip, start and end time, the course of a journey, and net-wide traffic state information are relevant factors. These requirements on a route analysis are accompanied by a realisable field observation. Thus, in the research project "wiki—impact of individual and collective ontrip traffic management on the traffic flow in urban networks"[1] different observation methods are combined:

- ANPR—automatic number plate recognition
- GPS-logging to observe the route choice behaviour
- interviews to explore the intention of the drivers
- driving simulation to get individual indications of driving and route choice behaviour.

The test field covered the northern part of Munich, including highways as well as urban streets with high traffic density and a lot of congestion.

1 ANPR: Automatic Number Plate Recognition

Overall 51 ANPR cameras had been installed on 22 control sections, so that 119 routes could be observed. The travel time was calculated for all routes in 5 min intervals. In addition a quality index, that indicates the number of recognized vehicles, was defined. As a further reference the ANPR routes and alternative routes were driven by observation cars in a field test. The analysis gave indications of data quality and was used as a scaling factor (average value and variance) and for calculation of the traffic state.

2 Revealed Preference Test

To analyse the route choice behaviour a combined method of observation of the real driven routes [Revealed Preference (RP)] and interviews how the test persons would behave in hypothetical route choice situations [Stated Preference (SP)] was chosen. An online form and the personal interview also gave insight in personal data of the test persons and their route knowledge. For the revealed preference study the comprehensive amount of 300 people had been tracked for 8 weeks by GPS-logs. The test persons were interviewed about their individual aspects and motivations. The study was split in three parts, each with 100 people and a duration of 10 weeks. The test persons had been commuters living in the northern surroundings of Munich and working in the northern city of Munich. The requirement was that they commute at least three times a week using their own car. So, the test persons had comparable knowledge on the infrastructure, route alternatives and on the typical traffic states.

3 Stated Preference Test

Three different traffic states had been displayed by four different media, on the one hand common display methods and on the other hand feasible future ones. The definition of the traffic states and the incident was done in that way that the main route was never the fastest and non of the alternative routes was always the fastest. To get realistic choice situations, the individual most used route and the second alternative for each test person of the field test was analysed. Based on that the four displayed routes had different traffic states and incidents. An incident was defined by a travel time increase (+5 min respectively +10 min), as an information of length in kilometres of traffic jam or as a coloured traffic state map with green, yellow, and red link information.

Fig. 1 Driving simulator (*left*) and cut-out of the simulated streets (*right*)

4 Driving Simulation Test

The test in the driving simulator was used to get information about subjective classification of traffic quality, also regarding the influence of familiarity with the network. Each of the 40 test persons had three rounds of simulation drives to absolve. On each drive of one kilometre street length they experienced different traffic states. Besides, two of the drives were shown a movie of the journey before to generate a kind of familiarity with and anticipation of the network and the traffic situation. The test persons were split into an experiment and a control group. The groups were shown different movies with varying traffic states. Based on that it was analysed how different anticipation on a trip causes different classification of traffic states (Fig. 1).

Comparing the rating of the test person for the total drive with the classification of each section, the influence of traffic states experienced before respectively the time on the subjective rating could be analysed. Therefore, for each section the test persons had to rate the traffic quality while driving. Directly after the simulator drive the test person had to rate the total trip giving a grade (six levels). Comparing the average rating of the sections with the total classification it is considerable, that the classification while and after the trip vary significantly. The rating after the trip was worse than while driving. This gives the conclusion that sections with a higher (worse) Level-of-Service are kept more clearly in mind. The tree-colour-rating (green, yellow, red) after the simulator drive in comparison to the ontrip rating confirms that. Therefore, the sector rating by a colour is mostly worse then said ontrip. The test persons had been faced to two different classification schemas: six level grades and three colours of Level-of-Service (LOS). It was not explained to the test persons that a grade corresponds to a specific colour, but the test persons had to do it by themselves. The result, as expected, is given in Fig. 2.

The different street links that the test persons drove in the simulator were assigned to different LOS. This theoretical distribution was compared to the

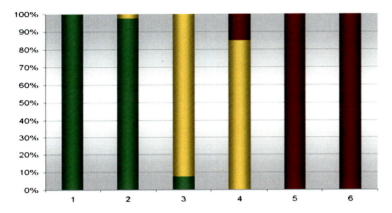

Fig. 2 Distribution of levels of service (LOS)

Table 1 Travel time factors

	Classification	Rating	Interview
Green	1	1	1
Yellow	2	1.67	2.27
Red	4	2.84	3.73

subjective rating of travel time of the test persons. Also the correlation of the travel time factor and a colour scaling was analysed.

Table 1 illustrates the results based on the test data. The test persons classify a street link on yellow already when travelling with a travel time of 1.67 while driving in the simulator. Thereby, they are not only more critical then they are thinking (2.27-times of travel time), but also as the standard scientific LOS classification (2-times of travel time). The deviation of the 'red link' rating by the test persons is even more obvious.

5 Analysis of the Route Choice Behaviour

The route choice behaviour is mainly influenced by the traffic state and the traffic management systems that the driver experiences during the trip. Thus, it is necessary to oppose the route choice data with the traffic and system conditions.

As a result of the study the assumption is that the anticipation plays a role on the strategic level, but not in the classification of the traffic state and of the individual travel times. That means that perception and strategic level have to be separated strictly.

Based on the experiment data of the test persons (logged trips and navigation use) and the objective traffic state and given traffic information, a statistic evaluation of the potential of traffic information systems was done. The focus was on the coherence of navigation use and LOS information and its influence on route choice.

In a first step the coherence of LOS and navigation use on the route quality was analysed. The quality indexes 'effective speed' and 'relative effective speed' were defined. They give a rate of the speed between origin and destination and are calculated on the length, the travel time and the detour factor of the route to the air-line distance.

Regarding all used routes the defined effective speed improved on ~ 7 % (1 m/s to13 m/s). But on trips with navigation use compared to these without, there was only an improvement of 3 %. In addition, the level of service information was used sometimes (<3 % of the logged trips) and got a higher improvement (10 %).

The relative effective speed is the ratio of effective speed and maximal speed on the origin–destination relation and gives a quality index of the used route. But the analysis showed that the quality of the used route could not be improved significantly using navigation and LOS information. But this is potentially caused by the traffic information (= basis for navigation and LOS) that is only available on motorways. The lack of information often causes detours on the minor street network. Besides it could not be analysed how often test persons used their private navigation systems on the recorded trips. It was calculated how big the travel time saving would have been under best conditions. The potential of detours calculated on travel time information for each driver of the test field would be 3 min on average or 9 % travel time savings. In 10 % of the trips the driver would save up to 75 % of their travel times. One of the main questions is, whether the drivers would change their favoured routes in case of traffic information of a potential travel time increase on the actual used route. And if a navigation or LOS request would support a detour. This would be an indication of willingness to use navigation systems. Several coefficients were analysed. The logistic regression analysis gives a model of impact of the factors familiarity with the route, relative effective speed, effective speed, and portion of congestion on the detour willingness. Notable is that all factors are significant even without a LOS request. But this could be caused by the test design that a lot of test persons have not only used the study navigation but also their private in car one. Another coefficient, that was tested, is the reliability of a route. It could be proved by logistic regression that the variance of the travel time has a significant impact on the route choice.

6 Conclusion

Today a lot of different traffic information systems are in use. In addition new telematic systems are under construction and testing in diverse research and development programs worldwide. The impact of the different management systems is only known partially. The evaluation of new and also existing systems is mostly incomplete or limited to quality statements. Thus, to fill the lack of knowledge a lot of empirical tests are done in the study 'wiki'. Especially the analysis of individual traffic behaviour was on focus.

The data of individual behaviour were related to the traffic state data. These data were used to analyse the coefficients on route choice. The personal characteristics were taken into account as well as the traffic state, route characteristics and the kind and quality of traffic information. A result is that the travel time has an important impact on the traffic choice. Besides, a strong main route preference of the test persons was evident. Alternative routes are only used if the travel time increase is up to a certain threshold. The focus was to analyse the impact of radio traffic services, variable traffic signs and navigation systems. A high impact on the route choice was radio traffic services and level of service information, displayed in the onboard navigation system.

Results of the driving simulator test are that reasons of delays have no significant impact on the traffic state evaluation by the test persons. But in contrast the experience of the traffic state on a street section influences the evaluation of the following section.

Reference

1. BMW AG et al (2012) wiki—Wirkungen von individueller und kollektiver ontrip Verkehrsbeeinflussung auf den Verkehr in Ballungsräumen. gemeinsamer Schlussbericht zum Forschungsprojekt 'wiki' gefördert durch das Bundesministerium für Wirtschaft und Technologie

Smart Automotive Apps: An Approach to Context-Driven Applications

Stephan Durach, Uwe Higgen and Michael Huebler

Abstract *Objective* Driven trough the new paradigm of application development for mobile devices, customers expect to have a device that can be updated with the latest and greatest features. Many useful applications have emerged from the existing ecosystem of SDKs, App Stores, etc. These applications are accessible anywhere, anytime, based on proper data connection. In the context of cars, one should be able to use apps without touching the mobile device. Cars offer a comprehensive infrastructure of control elements and screens optimized for usage whilst driving. Cars also offer a large amount of context knowledge from its sensor system. Leveraging the additional computing power and data connection of mobile devices in combination with the car enables a complete new level of features. *Methodology* An abstract interface between the car and the mobile device is implemented. The application itself is running on the mobile device and dynamically creates new features in the car. It is seamlessly integrated into the in-vehicle infotainment (IVI) system, which displays its features on the car screen. The application is controlled by the car interaction elements. By using this abstract interface the application on the mobile device has access to the sensor system and the audio system of the car. To avoid driver distraction guidelines and review procedures have been established. This is why applications can only connect to vehicles with proper authentication credentials. *Results* The IVI interface solves the objectives by offering (1) a "Remote HMI" interface that allows external applications to create their own HMI represented on the car display. User interaction is transferred to an external application through a notification service. (2) a "Car Data Server" interface sending vehicle sensor data to the mobile device. (3) an "Audio Service" playing audio in the car. Based on these technologies,

F2012-I03-010

S. Durach (✉) · U. Higgen · M. Huebler
BMW Group, Leipzig, Germany

new features such as MINIMALISM Analyser (a trainer for efficient driving) or Driving Excitement (a dynamic display of driving data, e.g. lateral/longitudinal acceleration, torque) are being developed. Furthermore, integrated versions of existing applications, e.g. music services like Pandora and MOG have been released. The IVI extension was introduced as an option for BMW Group vehicles in August 2010 (MINI)/March 2011 (BMW). At present, this option is compatible with the Apple iOS platform.

Keywords Apps · Car sensor interface · Feature upgrade · Mobile device integration · Human machine interface

1 Why Apps in the Car?

The app paradigm itself offers the unique opportunity to update consumer devices with the latest and greatest features at any point in time. In the context of cars we are faced with some additional challenges. On one hand there is a set hardware and software in the car that cannot be updated after its delivery to the customer. On the other hand we are living in a fast moving and ever-changing CE industry, which increases the computing power of their devices frequently. The user interface requirements in the car are completely different to a consumer device. Driving still remains the main task and any kind of distraction has to be avoided. A simple "copy and paste" of the device user interface (UI) is not appropriate—therefore consumer apps require the UI on the device as well as an additional UI for the car use case.

Introducing a new technology called Remote HMI [1] has enabled app developers to create apps that can be used whilst driving. It is possible to integrate exciting consumer apps and services such as Pandora, MOG, Aupeo seamlessly within BMW Group cars. However, this kind of integration is a re-packing of already existing services and features. The question is: "How can we leverage the specific information of our car sensors to create a new level of apps?" By introducing the concept of a so-called Car-Data-Server (CDS) we can expose such sensor data on the CE device through a defined interface.

Based on this information we can start to implement specific features of the car use case.

This technology is the answer to every customer's basic question: "How can I safely use my phones' exciting features and capability in my car?"

2 Approach

In order to integrate apps into the car, a number of interfaces need to be added to the IVI system:

1. The apps need to be accessible in the car whilst driving, i.e. they need to present a car-specific HMI on the car's screen, need to be controlled via the car's control elements (BMW iDrive controller, steering wheel buttons, etc.). For the extension of the car's HMI, the Remote-HMI (RHMI) interface is required.
2. The apps need to have access to sensor data from the car, e.g. its position or speed, to enable better car-specific use cases. The Car-Data-Server plays that role.
3. To play back audio for entertainment or information purposes, the Audio Service is required that allows for continuous audio playback or the insertion of short audio messages.

Several concepts have been evaluated for each of these interfaces [2]. The following technologies were chosen, because they allow for a very deep, premium-experience integration of external apps into the car.

2.1 Remote-HMI

For presenting the apps's user interface in the car, an HMI generation approach was selected. Within this setup, the smartphone app initially transmits a semantic description of its in-car HMI ("plug-in service") to the head unit. The head unit then interprets the description and lists the plug-in feature in the car's HMI menu tree at the requested position. The user can then navigate through the feature's HMI screens.

HMI elements ("widgets") that can be included in these screens contain e.g. scrollable lists of selectable items, labels, checkboxes, images and an icon-based toolbar.

Automatic reactions by the IVI on user interactions, such as opening a new screen when a toolbar icon is selected, can be specified in the HMI description. The smartphone app can also specifically register for notifications. Further app notifications include visibility of the feature in the car HMI or information about the previous status when the app was last connected.

In this case the app reacts appropriately and e.g. downloads contents from a backend server to then update its HMI elements in the car.

The initially transmitted HMI description consists of an XML-based model description as well as databases for texts and pictures. The databases are cryptographically signed, so that they can be used in production vehicles. This security mechanism ensures that only tested and reviewed HMI designs are released.

Table 1 Examples of CDS properties

Signal domain	Property	Content
Driving/engine	Odometer	Mileage (km)
	Speed actual	Current speed (km/h)
	Range	Expected fuel range (km)
Entertainment	Title/artist/album	Information about the current song (strings)
Navigation	Current position detailed info	City, country, street (strings)
	Next destination/final destination	Geographic position (latitude, longitude)
Sensors	Exterior temperature	Outside temperature
Settings	Language/units	Current user settings in the car: Selected language, km or miles, celsius or fahrenheit, etc

2.2 Car-Data-Server

To make car sensor data available to smartphone apps, a JSON-based abstraction layer was implemented in both the car's IVI system and the smartphone app framework (BMW AppKit). This abstraction layer decouples smartphone app development from car specifics like CAN or MOST message structures and data format. Instead, data structures (properties) with standard formats were defined, so that even app developers without a background in automotive software development can easily integrate them.

Smartphone apps can access CDS properties using a 'get' method, or they can subscribe to regular updates via a 'bind' method, which allows setting a desired update frequency (Table 1).

This approach also allows the abstraction from specific car models, because the CDS software can continue to offer the same interface to the smartphone app for new car models, even though they may employ a different data model on their internal bus system. However, such translation may not be possible if certain data becomes unavailable in future cars. On the other hand one may also like to introduce new properties in future versions. Therefore the CDS concept also includes versioning and methods for testing whether properties are available in the connected car.

2.3 Audio-Service

BMW Group cars typically handle two types of audio play back: 'Entertainment' audio is played back constantly until a different audio source is selected or the volume is muted. 'Interrupt' audio can be mixed with Entertainment audio for short messages such as turn-by-turn navigation announcements. Interrupt audio messages are subject to a prioritization scheme, in which safety-relevant audio signals rank the highest.

To make the functionality of the audio play back available to a smartphone app, an Audio-Service was created. The actual audio signal is transmitted to the car using existing channels (e.g. digital PCM audio on USB). However, to route the signal to the car's speaker system, the app needs to request one of three different channels: 'Entertainment', 'Interrupt', or 'Already Mixed'. Latter one is required since only one stereo signal path is available between the smartphone and the car. Moreover, the smartphone app would need to mix 'Interrupt' audio into an 'Entertainment' channel. This will not be allowed in situations where a car-internal high priority message is played. The interface also includes methods to test whether playback is currently allowed.

2.4 Other Considerations

A smartphone app using these interfaces can be developed independently of the automotive development cycles, which allows for a significantly shorter time-to-market. It can also be developed for cars that have been sold.

The abstract interface is the same in all BMW cars. This is why car specific testing is not required.

Smartphone apps in the car's IVI system need to fulfill the same driving safety requirements as 'native' apps. By using RHMI the developer can only access BMW approved widgets. This helps to avoid building apps that are not safe whilst driving. There are guidelines and review procedures for app developers in place. Authentication credentials are only issued after thorough reviews and can be revoked any time.

3 Example Application: ECO PRO ANALYSER

"What kind of apps can we build based on this technology?"—The ECO PRO ANALYSER is the perfect example. Introduced in 2012 by BMW, the new driving mode called ECO PRO selects predefined settings within the engine, optimises electric load and offers many other features for efficient driving. When switching to the ECO PRO mode in the central display of your car the ECO PRO screen shows up. The ECO PRO screen releases driving recommendations to optimize driving style.

The ECO PRO ANALYSER extents the ECO PRO mode with game design techniques. By introducing a leader board, star rating and visual feedback the app aims to keep the driver engaged to achieve the next level. The ECO PRO ANALYSER's goal is to promote a efficient driving style.

Fig. 1 IVI SYSTEM HMI screen with several apps

Fig. 2 Street metaphor nearly flat

Fig. 3 Example street metaphor of an inefficient driving style combined with efficiency recommendation

Smart Automotive Apps

Fig. 4 Evaluation

Fig. 5 Ranking shown on car display

The ECO PRO ANALYSER is fully integrated into BMW's iDRIVE system using the above-mentioned methods CDS and RHMI. The CDS has direct access to the driving mode ("sport, "comfort" and "ECO PRO"), tank level and distance (Fig. 1).

The status of efficiency is shown with five different street metaphors. A flat street represents the most efficient driving style whereas a wavy street the opposite. Figure 2 shows an example:

Fig. 6 Ranking shown on smartphone

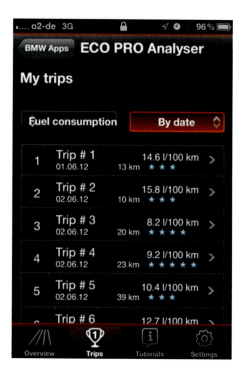

The driver receives immediate feedback if his driving style is inefficient. There is also a recommendation of how to improve efficiency, shown in Fig. 3.

There are two visual meters to indicate driver's efficiency and to encourage the driver to continue efficient driving. The visual meters indicate good efficient acceleration and anticipated driving. Depending on the number of stars, the street flattens. There is also a reward animation if the driver achieves additional stars in both categories.

Figure 4 shows the evaluation. There is detailed information for every trip as well as the average fuel consumption.

With the ranking list gives the driver an overview over his previous trips. These trips can be sorted by average fuel consumption or by date, see Figs. 5 and 6.

With this comparison (Figs. 5 and 6) the difference between the UI in the car display and on the mobile device itself becomes apparent.

4 Conclusion

We are convinced that the key to building a unique customer experience within the car is based on a specific set of methods and tools. The combination of the CE industry based apps together with the car specific data is a powerful solution. It is our aim to build a solution, which enables third party app developers to access the car. However, to accomplish seamless integration of the latest trends of external services a constantly updated IVI-system based on the Remote-HMI approach is required. In combination with the Car-Data-Server a new level of smart applications for the car will be implemented.

References

1. Hildisch A, Steurer J, Stolle R (2007) HMI generation for plug-in services from semantic descriptions. Proceedings international conference on software engineering (ICSE-2007), Workshop on software engineering for automotive systems, Minneapolis, MN, May 2007
2. Stolle R, Saad A, Weyl D, Wagner M (2007) Integrating CE-based applications into the automotive HMI. SAE technical paper 2007-01-0446, 2007, doi:10.4271/2007-01-0446

Pre- and Postdrive Predictions

Carsten Isert and Oliver Stamm

Abstract The prediction of a destination is of great use for advanced driver assistance systems (ADAS). However, clustering and predicting locations based only on latitude/longitude coordinates from cars lacks semantic information about locations and can be imprecise, especially once the car is parked. The functionality of these predictive systems can be greatly enhanced when the data used for predictions is extended beyond the vehicle and when predictions can already be made before entering the car. Therefore, we propose to include location-based services running on the driver's smartphone and provide the prediction via a web service. The inclusion of explicit check-ins is used to extract semantic meaning of places and to improve predictions beyond the parking spot of the vehicle. We implemented a prototype, which combines data from *foursquare* and *Google Latitude* and enables predicting not only the locations but also the most probable times when a user will arrive at a certain place and when that user will leave this place. This information enables preconditioning and optimization of charging strategies for electric vehicles and can improve recommendation systems. This paper demonstrates the overall system architecture and explains the prototype implementation.

Keywords Destination prediction · Time prediction · Location-based services · ADAS · Charging strategy

F2012-I03-013

C. Isert (✉) · O. Stamm
BMW Research and Technology, GmbH, Munich, Germany

1 Introduction

When predicting a car's or person's future destination and the respective temporal patterns it is not only relevant to know the rough position (latitude/longitude) of the destination, but much more useful to provide semantic meaning. Without semantic knowledge the system would not know that a location is called Sunshine Plaza, is a mall complex, and its correct street address. In this paper we define a location as a position with semantic meaning.

Examples of applications include recommendation systems that can plan activities based on future locations. If the next predicted location is a grocery store the user could be reminded to buy milk. Another scenario is the search for points of interest (POIs) near future destinations.

The contribution of this work is to predict destinations with semantic information. This is achieved by various improvements. To improve clustering accuracy, we extend prediction beyond vehicle usage by integrating the smartphone location-based services *foursquare* (www.foursquare.com) and *Google Latitude* (latitude.google.com). Furthermore, long term prediction is refined by mining temporal patterns from stops (period of time a user stays at a location). These goals are attained using three phases: data acquisition, clustering, and prediction.

2 Related Work

In order to make predictions based on a user's position data, it is necessary to derive a set of clusters from past locations. Ashbrook et al. [1] have found basic partition clustering methods like k-means to be too sensitive to provide reliable prediction. DBSCAN [2] shows sensitivity to its distance parameters. DJ-Cluster [3] improves on this with an enhanced density function that uses the notion of connectivity between neighbourhoods. However, those methods do not consider the temporal dimension and are not suitable for online learning. Alvares et al. [4] propose adding potential locations with semantic information to user traces to derive stops and movement behaviour from trajectories. The problem is that locations have to be predefined so that stops can be derived. CB-SMoT [5] tackles this problem by adding speed-density-based clustering to this approach using an absolute distance (Eps) to calculate the neighbourhood of each point. All of these approaches assume very accurate positioning data which is often not available.

After deriving a user's location history, the data can be used to predict future destinations. Laasonen [6] proposes to use approximate string matching. Froehlich et al. [7] compare algorithms for matching the current trajectory to a set of previously observed ones evaluating trip similarity using the physical distance of road segments. Miyashita et al. [8] suggest to focus on predicting only small parts of the route accurately by subdividing trajectories into intervals. BMW Research [9] implemented the iLeNa (intelligent Learning Navigation) system which uses recorded routes to predict sub-routes and destinations and generates an electronic horizon to optimize energy management in the vehicle. This system uses a variant of DJ-clustering based on the road network but is restricted to in car usage.

Kostov et al. [10] use time attributes to model transition probabilities using frequent pattern growth. In [11], Hidden Markov models are used to learn and predict destinations and trips. The fact that all attributes need to be encoded in intervals leads to significant effort for finding good interval sizes. Thus, fuzzy partitioning is proposed to distribute continuous states into several bins [12].

Many of the previously mentioned methods are specifically designed for traces of cars. Applying these methods to data from location-based services causes problems as LBS data is typically less accurate, incidental, and sparse. Additionally, our system combines different modalities and delivers more meaningful results.

3 System Overview

3.1 Data Aquisition

Quantity and quality of data are both important to obtain reliable results. At the same time, user tracking and storing of movement profiles has privacy implications. In our approach we can restrict predictions to data provided by explicitly the user (e.g. through check-ins) on different services and with the user's explicit consent. This section shows how to combine data from different services to make destination prediction applicable for daily life.

In-car navigation systems provide the possibility of storing and analyzing very precise data of a user's driving history. A vehicle-based prediction system can leverage the vehicle's ability to provide information about stops and its precise positioning on the road network. However, these technologies cease to work when the user leaves the car. We equipped two of our navigation research vehicles with hard- and software to automatically record this data. The basic system setup is shown in Fig. 1. The "route-recorder" stores the complete route information in a local database in the vehicle and additionally sends the stop position data to the backend where parking spot positions are stored. As an alternative we also provided a smartphone app where users could manually enter their parking spots so that data gathering was not restricted to people actually driving the research vehicles.

Intention and temporal patterns are misread if the user decides to use public transport instead of the car or uses the same parking spot with different intentions. A variety of smartphone location-based services can be utilized to close this gap. We selected *foursquare* as check-in based service due to its widespread use and easy to use API. Check-ins are used to specify the user's current location and usually contain name, position, and type of the place where the user is. Semantic information about locations is usually only given if the user actively checks in-which he might forget. Check-ins usually contain no information about the duration of the stay.

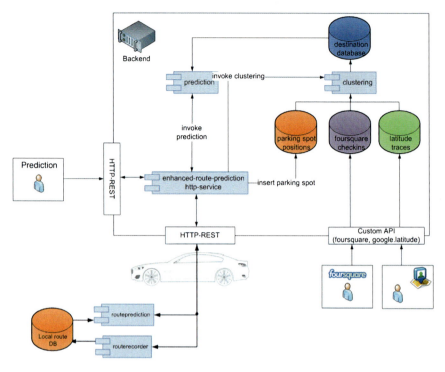

Fig. 1 Overview of the prediction system

As *foursquare* does not provide position traces of the user we added *Google Latitude* as additional service which logs the WGS84 coordinates of the smartphone. These traces strongly differ in accuracy and sparseness of positions. This leads to the conclusion that services need to be joined to obtain the required accuracy, quantity, and semantics. Hargreaves [13] presents research results on mobile check-in habits and analyzes trends of early adopters as well as mass consumers. He anticipates that the use of social location apps by mass consumers will continue to increase, which verifies the concept of combining different services. Our prediction backend service accesses the data from *foursquare* and *Google Latitude* via their official web APIs as can be seen in Fig. 1. This assumption is supported by Zickuhr [14] where almost a fifth of smartphone users used services like *foursquare*.

3.2 Service Combination

Figure 1 shows how the different data sources are transformed by our clustering algorithms into a coherent knowledge base to facilitate pattern analysis. We developed a density-time-based clustering algorithm that is based on CB-SMoT to enable the integration of the semantic data which is mainly from *foursquare*.

Candidate stops are a set of one or more geographically close locations to which a person has a personal connection and therefore is likely to go to. These stops are first extracted from the clustered data and joined in a kNN-manner to avoid overlapping candidate stops. Once the candidate stops are extracted, user traces are analyzed, combining a density-based and a spatio-temporal approach to find candidate stops at which the user has not checked in. Finally, stops at candidate stops are extracted. If positions are inaccurate, the algorithm might not be able to accurately determine the location of a user. Therefore, times of arrivals and times of departures need to be adjusted to eliminate inconsistencies. Additional information as parking positions from other services need to be associated with particular stops to add further semantic meaning. If a parking position cannot be assigned to a stop, a new candidate stop is created and clustered, leading to a new stop at the time of parking. The clustering is currently invoked manually via a web based interface as this is easier for research and demo purposes.

3.3 Prediction

The prediction algorithm is based on the destination database as shown in Fig. 1. With the current time and the current position as input the output is computed in each step as a list of next possible destinations with associated confidence values. Additionally, the times for leaving the current location and arriving at the respective destination are also provided with associated confidence values. The algorithm itself is a hybrid dynamic Bayesian network (HDBN) and works in three phases. In the first phase, the candidate stops to which the user went from his current location have to be extracted in the stop history. These candidate stops and the corresponding stops build the knowledge base for the prediction of the next location and temporal attributes. In the next phase, a transition function needs to be calculated for each candidate stop to be able to calculate confidence values for arriving at that candidate stop at a specific point in time. A similarity factor is calculated for each stop that describes the similarity to the current situation. This is based on the assumption that two similar stops also lead to a similar next location.

These factors can then be used to calculate the density function for the arrival time probabilities. As the peaks represent accumulations of historical stops, extracting the peaks leads to the most probable arrival times for each candidate stop. Eventually, the different candidate stops and corresponding arrival times are used to calculate the most probable departure time for each arrival time from the current location by calculating the similarity factors. However, this time they depend on the arrival times. The details of the clustering and prediction algorithm are not included due to space restrictions.

Fig. 2 Map view of predictions in the browser and list view on mobile phone

4 System Usage

The prediction can be invoked via a web interface and used from a website to access the information for user display or any other system like mobile assistants or the vehicle to enhance its prediction capabilities. A sample user interface can be seen in Fig. 2 for a general map based view that shows the current location of the user as a blue dot and the location of the car as a red parking sign on *Google Maps*. The suggested destinations are shown in the list view on a smartphone to the right. In this example it is Sunday evening 7:17 pm and the system predicts that the most probable destination is with a chance of 73.9 % with a departure time on Monday at 6:28 am arriving at the BMW Research and Technology office at 7:04 am. The other destinations are less likely and on Sunday evening. The user would reach these destinations not with the car.

Figure 3 shows the equivalent interface in the car. The confidence values are received from the backend and merged with the route data of the car. Figure 3 also shows the most likely route to the destination on a map based on OpenStreetmap (openstreetmap.org).

Combining this data enables relating parking spots to locations. Without this information only the parking positions could be supplied which might be meaningless to the user. Once the vehicle starts driving, the predictions within the car are updated frequently based on the actual route driven.

Fig. 3 Destination and route prediction in the vehicle

5 Results

We collected data for 6 weeks from 10 people with no restrictions made about the intensity a service had to be used. Overall, this resulted in about 3,500 check-ins, 140,000 latitude/longitude coordinates and 350 vehicle stops. Although no previous research can be directly compared to our results, a few of them can be used to get a general idea of the result quality.

Our approach is able to predict the correct location in the top 3 and the chance that the correct destination is within those is in more than 80 % of the cases after 20 days and about 100 trips.

This represents the steep learning curve of the location prediction algorithm. Additionally, the proposed location prediction algorithm predicts the correct category in the top 3 with a probability of 82.42 %. Froehlich et al. [7] mention that with their algorithm the correct route is within the top 8 matches on average while after one mile the algorithm has a prediction accuracy of 30 %. When using 80 % of the data as knowledge base, the correct location within the predicted ones is on rank 3.18 when using the location prediction algorithm proposed in this work. Additionally, when arriving at a location, the next location is correctly predicted in 55.21 % of the cases. Filev et al. [12] created an algorithm that, after 623 trips, is able to predict the correct destination within the top 3 candidates in 71.95 %. The location prediction approach proposed in this work shows that the correct location is in the top 3 predictions in 81.2 % of the cases.

These results show that the proposed prediction method returns good results compared to previous approaches.

Another important aspect of the prediction algorithm is the prediction of arrival and departure time. For all correct predictions of all users, the algorithm has an average error of one hour. However, this strongly differs from user to user. The most accurate result for one particular user is an average error of approximately 30 min.

Altogether, the location and time prediction algorithm achieves better results for the next n future locations, categories, and broad categories including arrival and departure times than comparable previous work.

6 Conclusion

We present a novel prediction method that extends the prediction from the vehicle to generally available information that can be used in several applications. The system is evaluated with simulations and a working vehicle prototype. It can be used to precondition the vehicle to the predicted driving scenarios. It addresses privacy issues as it can be configured to use only information that the user has explicitly shared via location-based services. Other location-based services can use the most likely places, e.g. enabling a POI search along the predicted route even after the car has been parked. Additionally, recommender systems can take advantage of this information to improve recommendations.

As the mobility patterns of individual users differ greatly, the algorithms should be able to adapt to these behaviours. Our test data is currently only available for 6 weeks. For better long-term evaluation it would be necessary to have data for at least one year.

References

1. Ashbrook D, Starner T (2002) Learning significant locations and predicting user movement with GPS. In: IEEE, Proceedings of the sixth international symposium on wearable computers, pp 101–108
2. Ester M, Kriegel H-P, Sander J, Xu X (1996) A density-based algorithm for discovering clusters in large spatial databases with noise. Computer 96:226–231
3. Zhou C, Bhatnagar N, Shekhar S, Terveen L (2007) Mining personally important places from GPS tracks. In: IEEE 23rd international conference on data engineering workshop, IEEE, pp 517–526
4. Alvares LO, Bogorny V, Kuijpers B, de Macedo JAF, Moelans B, Vaisman A (2007) A model for enriching trajectories with semantic geographical information. Proceedings of the 15th ACM international symposium on advances in geographic information systems. ACM Press, New York, p 1
5. Palma AT, Bogorny V, Kuijpers B, Alvares LO (2008) A clustering-based approach for discovering interesting places in trajectories. Proceedings of the 2008 ACM symposium on applied computing. ACM Press, New York, p 863
6. Laasonen K (2005) Clustering and prediction of mobile user routes from cellular data. In: IEEE Pervasive Computing, pp 569–576
7. Froehlich J, Krumm J (2008) Route prediction from trip observations. SAE International, Warrendale
8. Miyashita K, Terada T, Nishio S (2008) A map matching algorithm for car navigation systems that predict user destination. In: 22nd international conference on advanced information networking and applications—workshops (aina workshops 2008). IEEE, pp 1551–1556
9. Gruenweg T (2009) Big Navi is watching you. [Online]. Available: http://www.spiegel.de/auto/aktuell/0,1518,608617,00.html
10. Kostov V, Yoshioka M (2005) Travel destination prediction using frequent crossing pattern from driving history. In: Proceedings of the 2005 IEEE intelligent transportation systems, pp 970–977

11. Simmons R, Browning B, Sadekar V (2006) Learning to predict driver route and destination intent. In: IEEE intelligent transportation systems conference. pp 127–132
12. Filev D, Tseng F, Kristinsson J, McGee R (2011) Contextual onboard learning and prediction of vehicle destinations. In: IEEE symposium on computational intelligence in vehicles and transportation systems (CIVTS), pp 87–91
13. Hargreaves D (2011) The reality behind the "Check-In" hype
14. Zickuhr K (2012) Three-quarters of smartphone owners use location-based services. Pew Research Center Technical Report 2012 [Online]. Available: http://pewinternet.org/Reports/2012/Location-based-services.aspx

Part IV
Eco Driving Technology

Adaptive Cruise Control: A Behavioral Assessment of Following Traffic Participants Due to Energy Efficient Driving Strategies

Dirk Hülsebusch, Maike Salfeld, Yinchao Xia and Frank Gauterin

Abstract Improvement of vehicle safety and driving comfort have been the main objectives of driver assistance systems. Because of the general need to reduce the fuel consumption, the scope of driver assistance systems has been enlarged. Today, Adaptive Cruise Control (ACC) systems are developed which realize a more energy efficient driving style. However, energy optimal driving may lead to a trade-off between energy efficiency, vehicle safety and driving comfort. For example long coasting phases might irritate following drivers and provoke them to approach below the safety distance or overtake the ego-vehicle. This raises the question, how far energy optimized ACC affects following traffic and how the trade-off can be optimized to reach a higher acceptance. Within this work, a system to assess the driving behavior of following vehicles is developed and validated. The system is used to study the behavior of following traffic participants due to energy efficient ACC driving strategies. The results show that the median of the time gap between the following vehicle and ego-vehicle is significant lowered when driving with energy efficient ACC in comparison to driving manually. Whereas, the time to collision shows no significant difference between ACC and manually driven. At last, different concepts are presented which try to find an optimum between efficiency, safety and driving comfort.

Keywords Adaptive cruise control · ACC · Rear object selection · Driving behavior · Rear vehicle

---

F2012-I04-002

D. Hülsebusch (✉) · M. Salfeld
Robert Bosch GmbH, Munich, Germany

Y. Xia · F. Gauterin
Karlsruhe Institute of Technology (KIT), Institute of Vehicle System Technology, Karlsruhe, Germany

1 Introduction

Today, one of the main focuses in automotive industry is the reduction of vehicle CO_2 emissions and energy consumption. For reaching this goal, different strategies are pursued. On the one hand, the driving resistances are reduced. Important parameters are vehicle mass and air drag as well as the rolling resistance. On the other hand, the energy efficiency is increased by new powertrain technologies, like for example turbocharging, hybridization and full electric powertrains. Another alternative, which can be influenced by every driver, is the driving style. But the adaption of driving style is mainly limited by the willingness and ability of the driver. Therefore driver assistance systems are developed which support the driver to realize a more energy efficient operation and driving style. One well-known example is the gear shift indicator. The gear shift indicator displays whether the driver is using the most energy efficient gear according to the requested torque and actual engine speed [1]. In order to support the driver further, Adaptive Cruise Control (ACC) systems are extended with energy efficient driving strategies. One approach is to lower the accelerations and increase the time gap to the vehicle ahead. More sophisticated approaches use additional vehicle sensors and environmental information, e.g. navigation system, to drive energy optimal considering the speed limits and road slope ahead [2–4]. For example, anticipatory driving in front of a speed limit, in fuel cut-off mode, can lead to savings of about 16 % in comparison to an average driver [5]. It is comprehensible that energy efficient driving has an impact on the longitudinal vehicle dynamics. Therefore, it raises the question, if the change in driving style influences the behavior of other traffic participants significantly. A change in the behavior might lead to safety critical situations and acceptance problems with the driver assistance system. In existing studies such aspects of energy efficient driver assistance systems (Eco-ACC) regarding the vehicle behind the ego-vehicle are not considered in-depth. Within this study the driving behavior of following vehicles is analyzed when driving behind a manual and an Eco-ACC driven vehicle. In the first part, (see Chap. 2) a rear object selection for a backward looking radar is developed to track the driving behavior of the succeeding vehicle on the same lane behind the ego vehicle. The system is used for a proband study (see Chap. 3) in which a group of probands drives a defined course with and without Eco-ACC. For each proband and test drive the behavior of following vehicles is statistically analyzed. For example the time gap is compared for high and low traffic as well as for different road types, e.g. urban and extra urban like motorway and rural roads (see Chap. 4). In the last part different methods are presented to improve the conflict between energy efficient and safe driving (see Chap. 5).

2 Radar-Based Rear Object Selection

Within this chapter the developed object selection algorithm is explained. At first, the object selection is motivated by the means of an exemplary situation. Furthermore the architecture of the implemented algorithm is illustrated. Afterwards the object selection is validated on the basis of a test drive with different road types.

2.1 Object Selection Algorithm

For this study a single long range radar sensor of the 3rd generation from the Robert Bosch GmbH is used [6]. Advantages of radar based sensors in comparison to video sensors are for example the high range up to 250 m and the robustness against darkness and low visibility because of fog or rain. For analyzing the driving behavior of vehicles behind the ego-vehicle it is necessary to know which moving vehicle is the relevant object. Relevant are the vehicles of which the driving behavior is influenced by the ego vehicles motion. Therefore, in the following only the next vehicle driving on the same lane behind the ego-vehicle is considered. Naturally, the radar sensor detects objects as one point in the pane and consequently a situation analysis is necessary to determine whether the vehicle is moving on the same or neighbor's lane (see left part of Fig. 1). In order to assign an object to a lane, specifically the ego-vehicle lane, an algorithm is developed. For ACC systems, where a front sensor is used, for example standing objects like guardrails and moving objects in front of the ego-vehicle [7] as well as information about the ego-vehicle movement [8] are used to predict the future vehicle trajectory. In comparison to an ACC system, the vehicle trajectory has not to be predicted. In contrast, for an object selection behind the ego-vehicle, the vehicle trajectory history can be used. On the basis of the trajectory history (see left part of Fig. 1) the relevant vehicle can be determined and selected. The developed object selection algorithm [9] consists mainly of two parts (see right part of Fig. 1). As shown on the left side, the measurement data of detected radar objects are compensated in case of a misalignment of the sensor. In the next step, all irrelevant objects and disturbances are filtered. Irrelevant are standing objects, e.g. traffic signs or guardrails, and objects that move in the opposite direction as the ego-vehicle (see No. 4 and 5 in left part of Fig. 1). In parallel, the ego-vehicle trajectory history is calculated by the actual velocity and yaw rate. After that, the trajectory offset dh_n for every relevant object is determined. The trajectory offset dh_n is the lateral offset of the vehicle position to the trajectory history of the ego-vehicle (see No. 3 in left part of Fig. 1). With the trajectory offset and an assumption of the lane width, all relevant objects are assigned to the correspondent lane (see No. 1–3 in left part of Fig. 1). In the last step, the objects for the ego and the two neighboring lanes are sorted by their relative longitudinal distance dx_n in relation to the ego-vehicle position.

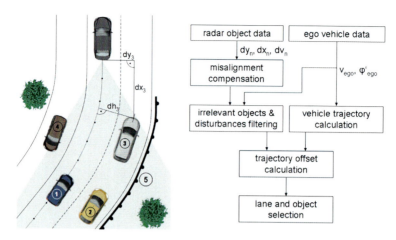

Fig. 1 Exemplary object selection situation (*left*) and flowchart of the object selection algorithm (*right*)

2.2 Object Selection Validation

The object selection algorithm is validated on a test drive around Stuttgart. The defined course has a distance of about 48.4 km and contains different urban, road and motorway sections as well as a varying amount of lanes [9]. A video sensor is used as reference that captures the situation behind the ego-vehicle. At first, the video data is used to label all situations with rear traffic. For example if a vehicle appears, the label "object appears" is set. If the object disappears, e.g. due to a lane change, the label "object disappears" is set. In the next step, the results of the object selection algorithm are compared to the labeled video data. This evaluation process is done automatically. Table 1 shows that in approximately 90 % of the driving time, the object selection of the algorithm is true. The true classification can be divided into two cases. The true positive rate indicates the ratio of time, the object selection detects a vehicle on the condition that a vehicle does exist. On the opposite, the true negative rate indicates the ratio that the object selection detects no vehicle under the condition that no vehicle is actual present on the lane behind the ego-vehicle. On all road types, the true negative rate is higher than the true positive rate (see Table 1). This might be explained by the fact that when an object is available, the object can be assigned to the wrong lane. Such an incorrect selection is not possible when no object exists. Overall, it can be noticed that the object selection performance with around 90 % is sufficient and can be applied to the study, which is introduced in the following part.

Adaptive Cruise Control

Table 1 Object selection classification rates for the next object on the same lane behind the ego-vehicle [9]

Classification rate	Total (%)	Urban (%)	Road (%)	Motorway (%)
True	91.67	93.18	92.27	89.28
True positive	89.50	91.01	87.68	89.24
True negative	95.49	96.99	97.57	89.37

3 Proband Study

In order to analyze the influence of Eco-ACC on the driving behavior of rear vehicles, an Eco-ACC proband study is carried out. Therefore the test vehicle is equipped with the back-ward looking radar system, which is outlined above. In the following, the study design and the used Eco-ACC test vehicle are described in more detail.

3.1 Study Design

The main objective of the proband study is to analyze how far energy optimized ACC can help the driver to reduce the fuel consumption under real traffic conditions. Thus, a sample of 30 probands is chosen that have to drive a defined circuit with and without Eco-ACC [10]. The age and gender is selected according to the German population to ensure a representative sample. Furthermore, the designed circuit around the city of Stuttgart is based on the German distribution of yearly driven road miles. The circuit (see Fig. 2) has a length of about 62.65 km of which 49 % on motorways and federal roads (red), 30 % on rural roads (green) and 21 % in urban surroundings (blue) [10]. As drivers require an adaption time to operate the system properly, every proband has to do a test drive to get familiar with the vehicle and the ACC system operation. After the adaption phase, the participants have to drive the defined circuit without and with ACC. Additionally, every proband has to drive during the rush hour and non-busy traffic period to consider also different traffic densities.

The manual and ACC drive are scheduled on the same weekday and time of day to minimize undesired traffic effects between ACC and manual test drives. Overall, every proband does 5 drives: an introduction with Eco-ACC, a manual in medium traffic, a manual in high traffic, an Eco-ACC in medium traffic and an Eco-ACC in high traffic. In order to even out influences on the test drives by adaption to the test vehicle, one half of the participants starts with the Eco-ACC and the other half with the manual test drive. Figure 2 shows exemplary the study sequence for the two probands P1 and P2.

Fig. 2 Proband circuit around Stuttgart (*left*) and exemplary sequence for two probands (*right*) [10]

3.2 Eco-ACC Test Vehicle

For the study, a standard BMW 530d is chosen. The vehicle is additionally equipped with a Bosch Stop & Go ACC system with a 77 GHz Long Range Radar, which is able to control the vehicle velocity and distance to a preceding vehicle between 0 and 200 km/h [11]. For analyzing the rear traffic, the backward looking radar sensor is installed. In comparison to a series ACC, the system is extended with an energy efficient driving program. The main changes related to the standard driving program are reduced accelerations and jerks as well as more usage of the fuel cut-off mode. The follow control is supplemented by a tolerance range instead of a fixed time gap. Thereby an inefficient driving style of the preceding vehicle is less copied. Additionally, the reaction on preceding vehicles with a high distance is decreased. Following a preceding vehicle, which is driving a stop & go cycle, leads to a reduction of 4.6 % related to the standard ACC. When the leading vehicle is driving a traffic jam cycle, an 11 % reduction is reached by using the Eco-ACC driving program [12]. The results of the proband study show a reduction of the fuel consumption within Stuttgart city during the main traffic time of about 3.5 % in comparison to the manual drive [10].

4 Results and Discussion

The following chapter describes the results of the driving behavior analysis of vehicles behind the ego-vehicle. At first, the procedure how the measurement data of the proband study is analyzed is explained. In the following sections the results of the distance, time gap and time to collision are presented and discussed.

Fig. 3 Analysis process of the proband study measurement data

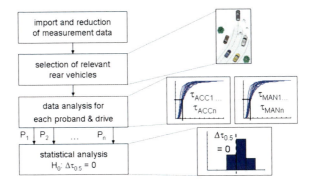

4.1 Data Analysis Process

In the first step, the proband study measurement data is imported and reduced to signals which are needed for the analysis (see Fig. 3). This data is fed to the object selection algorithm in order to find all relevant objects moving behind the ego-vehicle on the same lane. Afterwards, the selected object data is analyzed for each proband. For example, the distance dx to the following vehicle is determined by calculating the median distance over all time steps where a rear vehicle exists. The median distance is calculated for the manual and ACC drive of each proband. It is assumed that the manual and ACC drive of each probandare dependent and hence the Eco-ACC and manual data set form a paired sample, e.g.: (dx_{MAN1}, dx_{ACC1}) to $(dx_{MANn}; dx_{ACCn})$. On the basis of the n distance pairs, it is statistically analyzed whether the median distance of manual drives and Eco-ACC drives are equal, which is called the null hypothesis H_0. Therefore the difference distribution is calculated, for example $\Delta dx_1 = dx_{MAN1} - dx_{ACC1}, \ldots, \Delta dx_n = dx_{MANn} - dx_{ACCn}$, and it is tested if the median of the difference distribution has a significant deviation from $\Delta dx_{0.5} = 0$. In general, there are two kinds of statistical tests which can be used: parametric and nonparametric tests. In this case, a parametric t-test can not be used because the difference distribution is not normally distributed. Consequently, a non-parametric test is applied. The use of the non-parametric Wilcoxon signed-rank test for related samples is not possible due to the fact that the difference distribution is not symmetric [13]. Instead of the signed-rank test, a sign-test is applied, which does not require a symmetric distribution [13]. The sign-test rejects the null-hypothesis H_0 when the number of positive signs (w) of the difference distribution is too low. Beside the sign test, a confidence interval for the median of the difference distribution is determined [14]. The confidence interval is a measure of the tolerance and indicates in how many cases, e.g. 99 %, the calculated confidence interval covers the median of the difference distribution. In the following statistical analysis, a confidence level of 95 % is used. The above described analysis process is done for the distance dx, time gap τ and time to collision TTC regarding different road types and traffic densities. Due to measurement difficulties the data of only $n = 15$ probands could be recorded completely and thus be used for analysis.

Table 2 Results for the distance between ego and rear vehicle, medium traffic (*top*) high traffic (*bottom*)

Medium traffic	Total test drive		Urban		Road + Motorway	
	ACC	MAN	ACC	MAN	ACC	MAN
$dx_{0.5}$ [m]	23.099	27.541	16.407	18.065	31.922	33.023
$\Delta dx_{0.5}$ [m]	1.516		2.043		0.270	
p value (two sided)	0.607		0.118		1.000	
95 % confidence interval [m]	−0.205	6.528	−0.485	9.518	−4.835	7.287
High traffic						
$dx_{0.5}$ [m]	21.320	22.157	14.428	16.672	28.584	27.708
$\Delta dx_{0.5}$ [m]	0.602		0.376		−0.876	
p value (two sided)	0.607		1.000		1.000	
95 % confidence interval [m]	−1.947	4.030	−0.333	3.325	−4.478	2.034

4.2 Distance

The interaction process between two vehicles can be divided into a steady and a non-steady state. The steady state is characterized by a linked movement. This means that the rear vehicle is following the ego-vehicle with a low relative velocity, e.g. ±5 %, as well as comparable low distance. Especially during a car following phase, an adequate distance is important to prevent accidents in case of abrupt breaking of the ego-vehicle. Almost all determined distances during the steady-state are greater for manual (MAN) drives in comparison to Eco-ACC drives (see Table 2). Consider a level of $\alpha = 5$ %, the results are not significant, but a trend for medium urban traffic is observable. Moreover, the results show that the distance values in heavy traffic are slightly lower and closer together compared to medium traffic densities. As expected, the determined distances are lower when driving in the urban area than on the motorway. In summary, it can be stated that the distance between rear and ego-vehicle has a negative trend when driving with Eco-ACC in comparison to manual driving. But the distance is not the most accurate parameter to evaluate whether a safety issue arises. In addition, the actual velocity is needed to identify an unsafe situation.

4.3 Time Gap

A more meaningful parameter to analyze the space between two vehicles is the time gap. The time gap is defined as quotient of actual velocity and distance to the preceding vehicle [15]. For example, a time gap of 2 s and a velocity of 20 m/s equates a distance of 40 m. The results show that particularly in low city traffic the difference between Eco-ACC and manual driving is significant to the level 5 %

Adaptive Cruise Control

Table 3 Results for the time gap between ego and following vehicle, medium traffic (*top*) high traffic (*bottom*)

Medium traffic	Total test drive		Urban		Road + Motorway	
	ACC	MAN	ACC	MAN	ACC	MAN
$\tau_{0.5}$ [s]	1.469	1.582	1.497	1.681	1.343	1.391
$\Delta\tau_{0.5}$ [s]	0.113		0.271		−0.019	
p value (two sided)	0.118		0.035		0.607	
95 % confidence interval [s]	−0.126	0.467	0.000	0.497	−0.133	0.428
High traffic						
$\tau_{0.5}$ [s]	1.449	1.508	1.537	1.770	1.353	1.385
$\Delta\tau_{0.5}$ [s]	0.111		0.155		−0.014	
p value (two sided)	0.607		0.118		1.000	
95 % confidence interval [s]	−0.118	0.336	−0.021	0.536	−0.143	0.233

(see Table 3). Driving on road and motorway leads to a similar time gap and shows no difference between Eco-ACC and manual driving. Analog to the distance, in high traffic the deviation between Eco-ACC and manual driving are less than in medium traffic. In addition, the results for high traffic are not significant. Furthermore, it can be observed that on all road types and traffic densities the middle time gap falls below the German recommended value of 1.8 s. This means that using Eco-ACC results in a further unsafe distance behavior of vehicles following the ego-vehicle.

4.4 Time to Collision

Another parameter, which is often used to asses the criticality of vehicle approaching situations, is the time to collision (TTC). The TTC indicates the time until the vehicles will collide under the assumption that the actual driving state of the ego and following vehicle is maintained [15]. For example, a vehicle is approaching another vehicle with a relative velocity of 10 m/s and a distance of 50 m. In case that the relative velocity will be maintained, the vehicles will collide in 5 s. In contrast to the distance and time gap, the TTC is analyzed for all approaching situations between rear and ego-vehicle. Within this study, approaching situations are characterized by relative velocities lower than 5 % of the actual ego-velocity. Comparing the results for different road types, the TTC is lowest in high urban traffic (see Table 4). But the results show no significant differences between Eco-ACC and manual driving. Hence, the approaching behavior of following drivers is not changed when energy efficient ACC is applied in the ego-vehicle.

Table 4 Results for the time to collision between rear and ego-vehicle, medium traffic (*top*) high traffic (*bottom*)

Medium traffic	Total test drive		Urban		Road + Motorway	
	ACC	MAN	ACC	MAN	ACC	MAN
$TTC_{0.5}$ [s]	16.754	16.246	15.350	15.064	18.303	17.501
$\Delta TTC_{0.5}$ [s]	−0.140		−0.008		1.087	
p value (two sided)	1.000		1.000		1.000	
95 % confidence interval [s]	−2.536	1.926	−3.193	2.016	−1.484	2.957
High traffic						
$TTC_{0.5}$ [s]	17.090	15.513	14.121	14.186	18.969	17.747
$\Delta TTC_{0.5}$ [s]	−1.484		0.014		−1.594	
p value (two sided)	0.302		1.000		0.607	
95 % confidence interval [s]	−2.096	0.740	−1.585	1.486	−4.830	1.264

5 Concepts to Improve Trade-Off

The results of the proband study have shown that the time gap and distance to the following vehicle is decreased when the probands drive with Eco-ACC in comparison to manual driving. Within this study no further strategies like for example coasting in front of speed limits or curves are applied. Hence, it can be expected that using more sophisticated energy saving strategies would lead to a more distinctive change in driving behavior of following traffic participants. This might have the consequence that a trade-off between energy efficient and save as well as comfortable driving arises. Basically, there are two possibilities to handle this situation. On the one side, a driving style could be chosen that is less energy efficient with the disadvantage that the ego-vehicle consume more fuel than possible. On the other hand, an energy efficient driving style could be applied that ignores the fact that more safety critical situations could occur. In order to solve this conflict, new concepts are presented in the following. In many cases the reason for unsafe vehicle maneuvers is that rear traffic participants do not know the ego-vehicle driver intention. For example the situation might be defused by informing the rear traffic participants about the actual driving behavior. This could be supported by information about how much fuel the driver safe when he would accept the driving style. Though, this approach will not be applicable in every situation and with every type of following drivers. Another solution to improve the trade-off could be a cooperative approach. With the cooperative approach, similar to the platooning research activities [16], the drivers come to a compromise regarding the driving strategy. For example the rear driver accepts the ego vehicle driving strategy or they find a compromise between their intentions, e.g. efficient or time optimum. The information and cooperation concept have the shortcoming that for both concepts a car to car communication is required. Car to car communication is an active research field [17], but will not be available in short-term as well as in

Adaptive Cruise Control

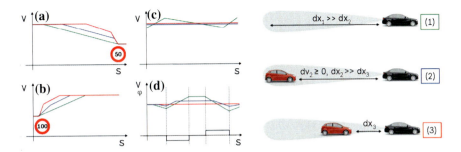

Fig. 4 Comparison of different driving strategies in dependency of the existence of rear vehicles

existing cars. For that reason a third approach named de-escalation is presented. Compared to the other concepts, a rear sensor is used to capture the scene behind the ego-vehicle. For example, if no vehicle is available (see Fig. 4 part 1) an energy efficient driving strategy is applied. In case of a speed limit, the distance of the coasting phase is maximized or the acceleration phase is stretched (see Fig. 4, part a and b, green line). Additionally, during constant speed sections or positive as well as negative road slopes (φ), the vehicle velocity can be varied to reduce the energy consumption (see Fig. 4, part c and d, green line).

When a rear vehicle is available or very close to the ego vehicle, the driving strategy is adapted to a more acceptable and less energy efficient driving style. Depending, for example, on the distance dx and relative velocity dv of the following vehicle, the driving strategy is modified to a more energy efficient or time optimized style (s. see Fig. 4, part a to d, blue and red line). All the above introduced approaches are possibilities to realize an energy efficient driving style while ensuring a safe and comfortable longitudinal vehicle control by adapting the driving style in dependency on the actual rear traffic situation.

6 Conclusion

The analysis of the proband study has shown that using energy efficient ACC in comparison to manual driving has an impact on the distance behavior of vehicles behind the ego-vehicle. Especially, in urban surroundings during low traffic densities following vehicles drive significantly closer to the ego-vehicle. In contrast, the time to collision has shown no significant difference between energy efficient ACC and manual driving. That might be due to the fact, that the applied energy efficient ACC strategy realizes only an energy efficient following control of preceding vehicles. For example, additional efficient strategies like coasting before speed limits, traffic lights and curves as well as variable speed trajectories are not considered within this study. Therefore further investigations are needed that clarify if differences between sophisticated energy saving strategies and normal driving strategies exist. In order to compensate potential conflicts while driving energy efficient, a concept has been shown that adapts the driving strategy with respect to the actual traffic situation behind the ego-vehicle.

References

1. van der Voort M, Dougherty MS, van Maarseveen M (2001) A prototype fuel-efficiency support tool. Transp Res Part C Emerg Technol 9(4):279–296
2. Dornieden B, Junge L, Themann P, Zlocki A (2011) Energy efficient longitudinal vehicle control based on analysis of driving situations. In: 20th aachen colloquium automobile and engine technology 2011, pp 1491–1511
3. Roth M, Frey M, Gauterin F, Lederer M, Radke T, Steinbrecher C (2011) Porsche innodrive—an innovative approach for the future of driving. In: 20th Aachen colloquium automobile and engine technology 2011, pp 1453–1467
4. Henn M, Lösche-ter Horst T, Schulze F, Bartsch P, Gadanecz A, Dornieden B, Junge L (2012) Energy efficient vehicle operation by intelligent longitudinal control and route planning. In: 12th Stuttgart international symposium automotive and engine technology pp 225–240
5. Reichart G, Friedmann S, Dorrer C, Rieker H, Drechsel E, Wermuth G (1998) Potentials of BMW driver assistance to improve fuel economy. FISITA World Automotive Congress, Paris
6. LRR3: 3rd generation long-range radar sensor (2009) Data sheet. Robert Bosch GmbH
7. Uhler W, Scherl M, Lichtenberg B (1990) Driving course prediction using distance sensor data. SAE technical paper 1999-01-1234
8. Moo S, Kang H-J, Yi K (2010) Multi-vehicle target selection for adaptive cruise control. Vehicle Sys Dyn Int J Vehicle Mech Mobil 48(11):1325–1343
9. Xia Y (2012) Development and validation of a rear object selection for novel driver assistance systems. Diploma Thesis, Karlsruhe Institute of Technology, Institute of Vehicle Systems Technology
10. Becker G, Kiefner D (2011) Design and validation of CO_2 reducing ACC control algorithm by the means of simulation and road test. EFA2014 internal final report, Robert Bosch GmbH
11. Adaptive Cruise Control: More comfortable driving, product information (2010) Robert Bosch GmbH
12. Schubert M (2011) Green ACC fuel consumption measurements (2011) Internal report, Robert Bosch GmbH
13. Duller C (2008) Einführung in die nichtparametrische Statistik mit SAS und R. Physica Verlag Heidelberg
14. Schlittgen R (2005) Das Statistiklabor Einführung und Benutzerhandbuch. Springer, Berlin
15. Vogel K (2003) A comparison of headway and time to collision as safety indicators. Accid Anal Prevent 35(3):427–433
16. Robinson T, Chan E, Coelingh E (2010) Operating platoons on public motorways: an introduction to the SARTRE platooning programme, 17th World congress on intelligent transport systems, Busan, Korea, Oct 25–29
17. CAR 2 CAR Communication Consortium Manifesto (2007) http://www.car-to-car.org/

System-Oriented Validation Aspects of a Driver Assistance System Based on an Accelerator-Force-Feedback-Pedal

Albert Albers, Alexander Schwarz, Christian Zingel, Jens Schroeter, Matthias Behrendt, Andreas Zell, Carmelo Leone and Antonio Arcati

Abstract Validation is considered to be the central activity in the product engineering process (PEP). It's the only way to check the fulfillment state of the customer objectives by the product. Accordingly, the product validation should apply real customer use case scenarios as test cases in terms of maneuver cycles. That means for the area of vehicle development that the continuous interaction of the systems driver, vehicle and environment has to be examined. For this purpose, the integrated IPEK X-in-the-Loop Framework has been developed, which exactly meets those requirements. It allows the application of objective-oriented, cost- and time-optimized systems engineering either for virtual systems (Model-in-the-Loop) or for real systems (i.e. Vehicle-in-the-Loop). To show the added value of system oriented validation and especially of the integrated IPEK X-in-the-Loop Framework for product development, the development process of a universal, energy-efficient Driver Assistance System (DAS) of Continental TEMIC is exemplarily applied. The challenge in the project at hand was to develop a cost-reduced, universal add-on fuel saving plug-in DAS for small- and medium-sized cars using the AFFP© as driver guidance system. It was asked to provide noteworthy fuel consumption advantage; extendable by additional interactions with other DAS's in order to easily increase its fuel-saving potential. The resulting, patented entire system, which allows up to 22 % fuel consumption reduction, depending on the driving situation and vehicle boundaries, is presented in this paper. Thereby, the main focus will be set on the validation of the development

F2012-I04-003

A. Albers · A. Schwarz · C. Zingel · J. Schroeter (✉) · M. Behrendt
IPEK— Institute of Product Engineering, Karlsruhe Institute of Technology (KIT), Munich, Germany

A. Zell · C. Leone · A. Arcati
Continental Automotive GmbH, Munich, Germany

activities, which have been performed by application of the IPEK-XiL Framework, e.g. at the IPEK Vehicle-in-the-Loop test bench.

Keywords X-in-the-Loop Framework (XiL) · Accelerator-Force-Feedback-Pedal (AFFP©) · Driver Assistance System (DAS) · Fuel Efficiency · Model-Based Systems Engineering (MBSE)

1 Introduction

Current and future trends in the automotive industry are normatively influenced by the global CO_2-emission problem and the dwindling oil reserves [1]. The main development challenge is to ensure high fuel efficiency and minimal emissions without compromises in key aspects such as driving comfort or performance.

A sustainable efficiency optimization of vehicles cannot be achieved only through mere optimization of the vehicle's components. Theoretical improvements on component-level must be validated in a realistic system environment, considering the interactions with the systems "driver" and "vehicle". Taking this into account, driver assistance systems can help to obtain the high aims of energy efficiency through guiding especially inexperienced drivers to drive their vehicle more efficiently [1].

Different approaches for driver assistance systems can help the driver to reduce fuel consumption. The most efficient of these systems are of a predicative nature. Unfortunately, such systems are very complex and therefore tend to be expensive. They are thus subject to the premium class [2]. Significantly cheaper systems, which can also be applied at compact- and mid-range vehicles, use rudimentary vehicle specific information in order to display general economic driving recommendations. But the achievable fuel savings from those systems are restricted [3].

Though, an effective, modular system, which is customizable for different vehicles using fuel-efficient driving maneuvers like coast and thrust, could have a significant impact on the total consumption of all vehicles because of the great number of vehicles in the vehicle classes mentioned above [4].

Validation is considered to be the central activity in the product engineering process (PEP). It's the only way to check the fulfillment state of the customer objectives by the product. This activity needs to be conducted continuously throughout the entire PEP [5]. Beginning with virtual validation by model-based simulation, wrong decisions can be avoided and the amount of necessary physical tests/prototypes can be minimized. This leads to significant reduction of cost-intensive product modifications in the late phases of the PEP and hence helps to save development time and costs. Accordingly, the product validation should apply real customer use case scenarios as test cases in terms of maneuver cycles. That means for the area of vehicle development that the continuous interaction of the systems driver, vehicle and environment has to be examined [1, 6].

System-Oriented Validation Aspects

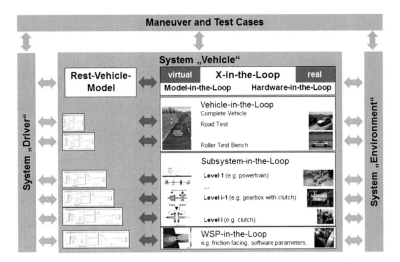

Fig. 1 The X-in-the-loop-framework with UUT as part of the system vehicle [6]

In the following, especially the continuous validation activities are going to be examined closely.

2 The IPEK X-in-the-Loop framework

The development of future vehicles comprises a lot of challenges because of the continuously increasing complexity. In order to successfully master these challenges, the X-in-the-loop approach has been developed, implemented and advanced at the IPEK, [6, 7]. This can also be used for validation purposes (Fig. 1).

The "X" represents the component to be examined, the so-called unit under test (UUT). This can be part of the system vehicle (e.g. a control device code, a complex powertrain or even the entire vehicle, see Fig. 1), the system environment (e.g. different frictional coefficients of the road surface), or the system driver. The UUT can exist in form of a real component or person as well as in form of a virtual simulation model. The framework is not limited to using one UUT. Several UUTs can be examined in parallel, which is especially important for the assessment of system and subsystem interactions [7]. In order to be able to close the loop between the interacting systems at all times, the remaining vehicle, environment and driver are simulated for every system integration level of the UUT. Only this way continuous, reproducible and transferable test results can be guaranteed [6]. Therefore, the XiL-framework is applied as methodical validation environment of the research project at hand.

3 Development Approach for Fuel-Efficient Driver Guidance Using the AFFP©

Continental TEMIC decided in 2009 to develop a fuel-saving driver assistance system in cooperation with the IPEK [8]. The IPEK contributed to the successful development through comprehensive validation activities using different levels of the previously introduced X-in-the-Loop validation framework. The paper at hand focusses on elucidating the application of the different XiL-environments and the benefit for the product development. For that reason, the development process of the modular plug-in DAS will only be described briefly using the implementation example of a medium-sized vehicle with ACC.

3.1 System of Objectives (Excerpt)

The core requirements for the plug-in DAS are as follows:

- The system must be adaptable to vehicle-specific operation states, in order to ensure maximal fuel efficiency. This adaption must be cheap and easily to apply.
- The human machine interface (HMI) of the system must be universally applicable and adjustable to the target customer in order to ensure maximum acceptance. Furthermore the system must be able to be overruled and/or deactivated at all times, since some situations require full driver control.
- The complete system must be capable to easily being integrated into the target vehicle (plug-in), especially with regard to the software integration.
- The system should be expandable towards a predictive system by integrating additional sensors/information.

3.2 Human Machine Interface

In modern vehicles the driver must process a lot of information (i.e. traffic situation, operating conditions and settings of the vehicle), prioritize and interpret them in order to derive the respective actions and operations. Most of the information appear optical or acoustical. It becomes apparent that the drafted amount of information will increase due to the use of additional systems, which deliver various information of diverse importance [9]. Therefore it is inevitable to develop new communication and interaction concepts for security relevant or fuel consumption reducing DASs.

Scientific studies have shown that the tactile sense provides promising capabilities to transmit such special information [10]. Commercial applications have already proven this [3]. The Accelerator-Force-Feedback-Pedal (AFFP©) developed by Continental TEMIC allows haptic stimulation of the driver's right foot

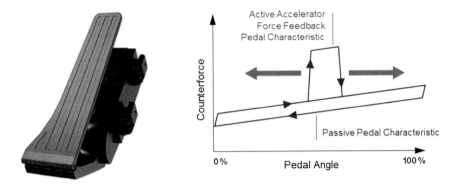

Fig. 2 Accelerator force feedback pedal (AFFP©) by continental TEMIC (*left*). Possible counter force paths of the AFFP© (*right*)

(Fig. 2). The AFFP© uses an integrated control unit which controls an electric motor. The motor generates a variable counter force, which can be used to assist the driver in finding an efficient pedal position. A jitter function can signalize gear changes or emerging dangerous situations (Fig. 2).

The pedal control unit can additionally integrate the code of the new energy efficiency DAS algorithm, wherefore existing control units don't have to be modified. Hence the AFFP©-based DAS can be realized as an autarkic plug-in system. The system integration is thus independent from the development process or the lifecycle of the vehicle and it's EE-architecture. This leads to the decision to implement the new DAS on the basis of the AFFP©. In general, the procedure described in the following is universally valid.

3.3 Identification of Saving Potentials

Firstly, an extensive collection of possible driving situations (maneuvers) with realizable energy-efficient driver guidance, optionally using additional information from existing sensors, is acquired. Based on the target-group-specific properties of the system "driver" (e.g. typical acceleration behavior [11]), narrowed down by the system "environment" (e.g. maximum speed, expectations of other road users regarding driving behavior), a list of maneuvers with saving potential that are acceptable for the driver results. The maneuvers consist of the basic maneuvers "acceleration", "deceleration" (including coast and thrust) and "constant speed" (analogous to [12]) and derived combinations of these basic maneuvers.

From this list, the maneuvers with high fuel saving potential are chosen. The quantity of fuel-saving potential is evaluated by simulation. For this purpose, a vehicle model has to be used, e.g. by means of AVL Cruise. Based on experience a model with six-cylinder engine and automatic transmission was applied and parameterized within the following validation steps.

The following maneuvers are going to be discussed in more detail as a representation for all efficient maneuvers. Minimum required sensor systems for detecting the traffic conditions (i.e. speed limits) are stated in brackets.

- Energy-efficient *acceleration* independently from target speed (no sensors required):
- *Follow to stop* (1,500 m): acceleration from 0 to 100 km/h, followed by section at constant speed and finally a deceleration to 0 km/h. The complete maneuver sequence is limited to a distance of 1,500 m (ACC required).
- *Temporary speed limit*, no vehicle ahead: the initial velocity is 100 km/h. After 1,000 m, a speed limit of 50 km/h (i.e. a town) appears. The speed limit takes effect for a distance of 1,000 m. Finally, the vehicle is accelerated to 100 km/h again. The total length of this maneuver sequence is 2,800 m (GPS required).

3.4 Realization of the Software Code

After having identified maneuvers with high fuel-saving potential and the according physical regularities, the operating strategy for the AFFP© control unit is developed. For the first concept of this code, the relevant physical structure, the software structure and according interfaces is initially modeled in SysML (see left side in Fig. 5). The purpose of the application of a discipline-crossing system model is to obtain a better system understanding and a model-based and hence consistent, redundancy-free documentation of the system architecture. Furthermore, the involved components (i.e. the engine control unit) and relevant characteristic parameters, curves or maps could be set into relation and be graphically visualized. Based on those gathered information, system states and performed activities within the control unit are modeled in state diagrams and activity diagrams (right side in Fig. 3).

An example for identified software states is the driver guidance during cruise control at a constant speed. When the current vehicle speed decreases below a lower boundary value (i.e. 2 km/h less than set speed), the AFFP© shall switch to the state "speed is too low". In this case, the activity relocate counterforce position towards higher pedal position", i.e. from 50 to 55 % pedal position is performed. Hence, the states and the conditions for the transition between them is modeled in state diagrams and the performed activity steps with according signal flows are modeled in activity diagrams. The system model is used as concept model for the prototypic implementation in Simulink, which is elucidated in the next paragraph.

3.5 Creation of a Simulink Model of the Fuel-Save-Algorithm

The next step consists of the creation of a Simulink model of the fuel-save-algorithm with defined interfaces to the hitherto existing controller code of the

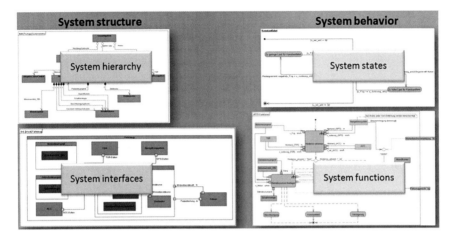

Fig. 3 System architecture and concept for control unit software code

AFFP©. There are four controllers which calculate the position of the gas pedal at all times and communicate the position to the driver with help of the actuator. An implemented algorithm chooses, which controller to use actively in the current driving situation. Characteristic parameters, curves and maps for the different controllers have to be determined by measurements of the test vehicle.

For optimal driver guidance, the AFFP© system must know the fuel consumption of the vehicle in each engine operation state in order to guide the driver within an optimal area. Hence, the characteristic map of the vehicle has to be parameterized. This was not known for the applied test vehicle, wherefore it had to be identified by measurements. Additionally, the test vehicle was equipped with automatic transmission, so the gear shift points also had to be determined by measurements. The vehicle map was identified on a "static" way in order to minimize influences from dynamic effects. This means, that a grid of operation points to measure was defined. The areas between the measured points will later be interpolated by the control unit (see upper left diagram in Fig. 4). Coevally, the gear shift points were identified on a "dynamic" way by conducting accelerations with constant acc. pedal positions and recording the speeds, where gearshifts were performed by the vehicle. That information was afterwards integrated in the characteristic map for the optimizer (see Fig. 4).

3.6 Counter Force Path for Driver Guidance Via AFFP©

There are different possibilities of modulating the counter force in order to guide the driver via AFFP© to adequately carry out the respective maneuver. Preliminary in-house assessments by Continental TEMIC have shown that in this case it is optimal for the driver to be able to "feel" the best pedal position at all times (all speeds) due

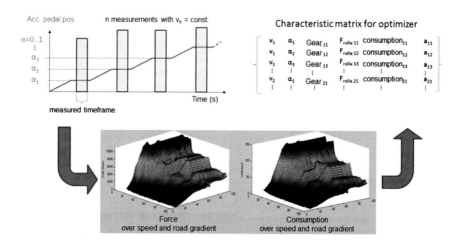

Fig. 4 Identification method for fuel consumption map

to an increased counter force. The modulated counter force "moves" continuously along the range of pedal positions. The driver only has to step on the gas and let the counter force guide him. The driver can override the counterforce at all times, simply by actuating or releasing the accelerator pedal, should the situation require it.

4 System Validation

The required validation environments and the respective validation activities will be illustrated more closely in the following chapters. It should be noted that those environments were also used for the development in identical or similar configuration. This similarity also reduces development time.

5 Virtual Validation: Code-in-the-Loop

The virtual code validation environment was established in Simulink. The environment model consists of two main components: the test environment and the model of the remaining vehicle. The fuel-save algorithm calculates a recommended pedal position for the current driving situation, based on the maneuver cycle information from its virtual environment using sensor signals from the vehicle CAN-Bus. The pedal position is transmitted to the vehicle model. The replied signals, calculated from the reaction of the vehicle model, are set up identically with those available in a real vehicle (Fig. 5).

The algorithm calculates the optimal pedal position using four controllers. Logic decides through continuous comparison of distance to the next obstacle and

System-Oriented Validation Aspects

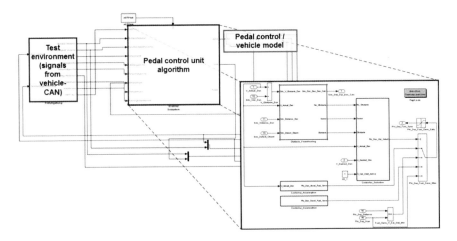

Fig. 5 Code-in-the-loop validation environment

energy efficient deceleration curves, which of these controllers must be activated. The established virtual test environment in the Simulink model facilitated a first validation of the controller logics with very low efforts. Furthermore, the model could easily be transformed to executable code through implementation of reality-conform signals.

6 Augmented Validation: HMI-in-the-Loop

In this validation step the driver guidance system, which was already developed by Continental TEMIC, has been validated successfully. The driver can follow the given driving instructions very well. This HMI validation has to be carried out with a real driver interacting with the real AFFP©, since the interaction complexity could not be virtualized with acceptable effort. The remaining vehicle and the environment in which the driving scenarios take place were simulated (Fig. 6).

Thus the validation of the HMI could be reproducibly realized in a driving simulator, without the need of an expensive test vehicle. The meanwhile advanced validation environment is described in [9].

7 Real Test Validation: AFFP©-System-in-the-Loop

After having validated the interaction between the AFFP© and the driver, the identified settings and characteristic curves have been implemented in the control unit source code by Continental TEMIC. The entire AFFP© System was then installed in a test vehicle in order to validate the system operation under realistic

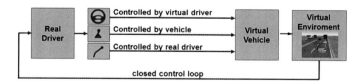

Fig. 6 XiL-Setup for HMI-in-the-loop test environment (schematic)

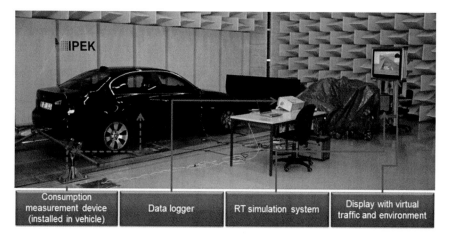

Fig. 7 AFFP©-in-the-loop test environment on roller test bench

conditions. Therefore, the validation experiments were conducted on a four-wheel roller test bench as shown in Fig. 7.

For fuel consumption logging a fuel consumption measurement device is installed in the vehicle. A data logger aligns this signal with the coevally recorded CAN-data (the operating conditions of the vehicle like speed, engine speed etc.). The real-time simulation system is responsible for the rest-vehicle simulation as well as the simulation of traffic and environment (i.e. course of the road). A display was used to visualize the virtual track for a test driver in order to simulate realistic traffic situations like closing up to slower vehicles driving ahead.

The applied test vehicle was equipped with automatic transmission. The shifting strategy was not manipulated here due to missing access on the CUs. The assessed maneuvers were all of longitudinal dynamic nature, therefore steering was not necessary. More information about the hardware and software setup of the applied vehicle-in-the-loop test bench can be found in [6].

8 Fuel-Saving Results of AFFP© Driver Assitance System

The results presented in this chapter were captured with the realized DAS using the AFFP©, applied in a real test vehicle on the roller test bench.

System-Oriented Validation Aspects

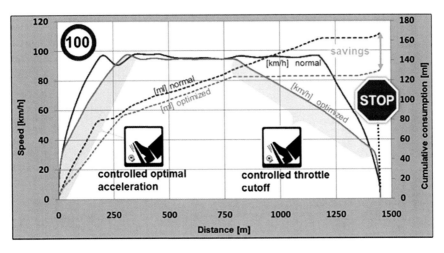

Fig. 8 Results from Follow to stop with/without DAS (1,500 m)

Follow to stop (1,500 m) In this maneuver sequence, the driver accelerates the vehicle from standstill to 100 km/h, drives constantly at 100 km/h and finally stops the vehicle after approximately 1,500 m (Fig. 8). The driver from the presented example is very sporty and accelerates faster than the optimized driver, who follows the guidance of the AFFP©. This is followed by the constant speed section. The normal driver starts decelerating relatively late, about 300 m before the stop. The optimized driver is guided to decelerate much earlier in order to activate fuel cut-off at motoring by the AFFP©.

Noticeable in the shown example is the fact that only the acceleration procedure already results in a fuel saving of 5 %. The fuel cut-off in motoring mode is the main reason for the resulting, significant fuel savings of 22.3 %.

Temporary Speed Limit This maneuver describes a typical traffic situation: the vehicle drives with 100 km/h on a rural road and approaches a village with a speed limit of 50 km/h (Fig. 9).

A normal driver would start to decelerate lately and use the mechanic brake. The optimized driver is instructed to fully release the gas pedal much earlier without using the brake pedal in order to activate fuel cut-off for at motoring much earlier. In the next step both drivers pass the 1,000 m long village with 50 km/h. Afterwards they can accelerate again. The normal driver intuitively accelerates relatively slow, which is not energetically reasonable. The optimized driver follows the instructions of the AFFP©, which provides an optimal acceleration curve.

The most significant fuel savings occur as expected during the deceleration with fuel cut-off. A relatively long distance is covered without fuel consumption in contrast to the normal driver, who decelerates much later and uses fuel cut-off only for a short time. With the AFFP©-fuel-save-function, a total fuel saving of 9.9 % results for this maneuver sequence.

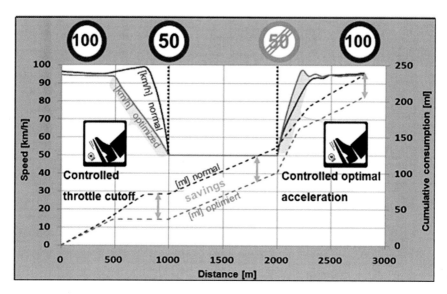

Fig. 9 Temporary speed limit

9 Summary and Outlook

In this paper the development and validation of a cost-reduced, universal fuel saving plug-in DAS for small- and medium-sized cars using the AFFP© as driver guidance system was presented. It was asked to provide noteworthy fuel consumption advantage, extendable by additional interactions with other DAS's or considering additional sensor systems for more comprehensive prediction of traffic situations in order to easily increase its fuel-saving potential.

For time efficient and cost effective development, three validation steps with specific validation environments based on the XiL-Framework were implemented and used for the validation of the DAS at hand: Code-in-the-Loop, HMI-in-the-Loop and AFFP©-SYSTEM-in-the-Loop. Thereby the continuous interaction of the systems driver, vehicle and environment was examined. The results demonstrated the effectiveness of the methodical proceeding: fuel consumption could be reduced by up to 22 %, depending on the applied maneuver sequence.

In order to reduce development and measurement time on the roller test bench, meanwhile the described methods were advanced by the IPEK [7]. At the moment a Reinforcement Learning Algorithm [1] is being linked with the vehicle on the roller test bench and the virtual environment. It will optimize online e.g. the acceleration maneuver. Afterwards the results can be immediately flashed onto the CU and validated on the roller test bench by a test driver. This significantly reduces development and validation time.

References

1. Albers A, Schroeter J, Frietsch M, Sommer H (2011) New approach for computation of predictive, fuel efficient vehicle operation strategies based on a self-learning algorithm. In: Proceedings of 56th scientific colloquium
2. Roth M, Radke T, Lederer M, Gauterin F, Frey M, Steinbrecher Chr, Schröter J, Goslar M (2011) Porsche InnoDrive—an innovative approach for the future of driving. In: Proceedings of 20th Aachen colloquium automobile and engine technology
3. Nissan News (2008) World first eco pedal helps reduce fuel consumption. http://www.nissanglobal.com. 04 Aug 2008
4. Raubitschek Chr, Deuble P, Baeker B (2011) Approach to energetic evaluation of different decelaration methods by the use of predictive information. In: Proceedings of energy efficient vehicle technology I, pp 144–156
5. Albers A (2010) Five hypotheses about engineering processes and their consquences. In: Proceedings of TMCE
6. Albers A, Düser T (2010) Implementation of a vehicle-in-the-loop development and validation platform. In: Proceedings of the FISITA 2010
7. Albers A, Schröter J, Behrendt M (2012) Development environment for automated configuration and generation of optimized, real-time capable for fuel-consumption-vehicle-models. In: Proceedings of 5th IAV simulation and test berlin
8. Patent: Vorrichtung zur Erzeugung einer zursäztlichen Rückstellkraft am Gaspedal und Verfahren zu deren Betrieb (2010) Patent number DE 10 2010 031 080 A1
9. Albers A, Schröter J, Düser T (2009) Integrated validation environment for HMI-testing of novel DAS. In: Proceeding of VDI testing and simulation
10. Zell A (2010) Active accelerator pedal as driver interface. In: ATZ 2010, vol 112
11. Demel B (2007) Contribution to predict driving styls. VDI-Report 2007, vol 2015
12. Neunzig D (2002) Potentials of predictive driver assistance regarding fuel consumption reduction. In: Proceeding of 11th Aachener colloquium on vehicle and engine techniques

Part V
Harmonization and Regulation of ITS Systems

Dynamic Optimal Model Researches for Correlative Intersections Control Based on Particle Swarm Optimization Aiming at Multi-Subsystem

Yifeng Huang and Song Luo

Abstract In this thesis, a dynamic optimal model of correlative intersections control aiming at Multi-subsystem, which based on particle swarm optimization have already been given, aiming at the delay of vehicles stopping. And as an application of this thesis, connecting this model and particle swarm optimization aiming at multi-subsystem to have an operation that use the north Tian he area of Guangzhou city as a example, have also been showed. The result of the operation and the simulation test all shows that this model act better than the custom ones. The high efficiency is in favor of building up the real time area traffic control system.

Keywords Optimal control model aiming at multi-subsystem · Particle swarm optimization · Delay of vehicles stopping · Simulation test

1 Preface

To control single intersection always can't reach the best effect when we want to make a network control, even it reaches the optimal control effect, so the traffic signal control system that make the correlative grouped intersections as the control object appeared, and many researches have been done by researchers abroad [1–5].

F2012-I05-002

Y. Huang (✉)
Guangzhou Automobile Industry Group CO.,LTD, 510030, Guzngdong, China

S. Luo
South China University of Technology, 510640, Guzngdong, China

2 Proposition and Deduction of the WH Traffic Flow Discrete Model

The WH traffic flow discrete model proposed in this article can be expressed by the flowing words:

The arrival rate of the vehicle on a downstream section have a relationship with the pass rate of the vehicle on upperstream stop line section:

$$q_d(i + \frac{\beta l_i}{\overline{v_i}}) = Fq_o(i) + (1-F)q_d(i + \frac{\beta l_i}{\overline{v_i}} - 1) \tag{1}$$

In this equation: $q_d(i + \frac{\beta l_i}{\overline{v_i}})$ denotes the expected arrival rate of the vehicle on a downstream section during the no.(i + t) period of time; $q_o(i)$ denotes the pass rate of the vehicle on a upperstream stop line section during the No.(i) period of time; $\overline{v_i}$ is the average speed of the traffic flow during it flow from the upperstream section to the downstream section. $G = \frac{\beta l_i}{\overline{v_i}}$ denotes the average time that the first vehicle in the traffic flow which have been stopped by the red light need to take to reach the downstream section (also can be regarded as the least time that the vehicle in one of this kind of traffic flows must take), and this must be the measured value; if there are no measured value, and the gradient of the road between the two section is 0, take $\beta = 0.8$. F denotes the coefficient which describe the dispersion of the moving traffic flow, called traffic flow discrete coefficient, the F actually reflect the different between the speed of each vehicle in the traffic flow.

The deduction of this model is:

Set $i = j - \frac{\beta l_i}{\overline{v_i}}$, so $q_d(j) = Fq_o(j - \frac{\beta l_i}{\overline{v_i}}) + (1-F)q_d(j-1)$, and also we can get
$q_d(j-1) = Fq_o(j-1-\frac{\beta l_i}{\overline{v_i}}) + (1-F)q_d(j-2)$ so:
$q_d(j) = Fq_o(j - \frac{\beta l_i}{\overline{v_i}}) + F(1-F)q_o(j - \frac{\beta l_i}{\overline{v_i}} - 1) + (1-F)^2 q_d(j-2),$ and
$q_d(j-2) = Fq_o(j-2-\frac{\beta l_i}{\overline{v_i}}) + (1-F)q_d(j-3)$, now we can substitute this two equation into the above expression, results in that:

$$q_d(j) = F(1-F)^0 q_o(j - \frac{\beta l_i}{\overline{v_i}}) + F(1-F)^1 q_o(j - \frac{\beta l_i}{\overline{v_i}} - 1) + F(1-F)^2 q_o(j - \frac{\beta l_i}{\overline{v_i}} - 2) + (1+F)^3 q_o(j - \frac{\beta l_i}{\overline{v_i}} - 3) + \ldots\ldots + F(1-F)^{j-t-1} q_0(1) \tag{2}$$

In the same way we make the substitution again and again until $q_o(1)$ has been substituted.

As $q_o(0) = 0$, so $q_d(j)$ can be make like that:

$$q_d(j) = F(1-F)^0 q_o(j - \frac{\beta l_i}{\overline{v_i}}) + F(1-F)^1 q_o(j - \frac{\beta l_i}{\overline{v_i}} - 1) + F(1-F)^2 q_o(j - \frac{\beta l_i}{\overline{v_i}} - 2) + (1+F)^3 q_0(j - \frac{\beta l_i}{\overline{v_i}} - 3) + \ldots\ldots + F(1-F)^{j-\frac{\beta l_i}{\overline{v_i}} - 1} q_0(1)$$

Accordingly,

$$q_d(j) = \sum_{i=1}^{j-\frac{\beta l_j}{v_i}} q_0(i) F(1-F)^{(j-\frac{\beta l_j}{v_i}-i)} \qquad (3)$$

3 Optimization Based on Particle Swarm Algorithm

Let M denotes the particle swarm scale, the dimension of particle i(i = 1,2,....M) is D, and the position of the No.i particle in D-dimension is denoted as Xi = [x i1, x i2...xiD], flying speed could be denoted as vi = [vi1, vi2...viD]. According to the fitness function, the No.i particle can find its optimal location which can be found at present, called the individual extremum 'pbestid'. Each particle also knows the whole swarm's optimal location which can be found at present, called the global extremum 'gbestid'. During each time of searching, particles update their speed and position through track the pbestid and gbestid. The equation of the updating speed and position can be described like the following:

$$v_{id}^{k+1} = w v_{id}^k + u_1 r_1^k (pbest_{id}^k - x_{id}^k) + u_2 r_2^k (gbest_d^k - x_{id}^k) \qquad (4)$$

$$v_{id}^{k+1} = x_{id}^{k+1} - x_{id}^k \qquad (5)$$

$$w = w_{max} - K \frac{w_{max} - w_{min}}{\theta} \qquad (6)$$

In the equations: d = 1,2,....,D; v_{id}^k, x_{id}^k denote, respectively, the speed and location of the particle i during the No.k searching, k∈(1, H); H denotes the maximum iterative searching times; u_1 is the cognitive factor; u_2 is the social factor; r_1, r_2 denote respectively a random number range from 0 to 1; w is the inertial factor, w∈[w_{min}, w_{max}]; when the value of w is big, for PSO, the global searching ability is good, when it is small, the local searching ability is good.

With the method of inertia weight particle swarm algorithm, the real-time control of the correlative grouped intersections signals can comply the following procedure:

1. Ascertain parameters in the algorithm : particle swarm scale M, dimension of particle D, particle motion range [x_{min}, x_{max}], and speed range [v_{min}, v_{max}], inertial factor w, cognitive factor u_1, social factor u_2, maximum iterative searching times H;
2. Randomly generate a particle swarm, the scale of the swarm is P, initialize the flying speed and position of the particle swarm;
3. Use the measure function of location optimization, which corresponding to the fitness function, to calculate the fitness of every particle;
4. Find the pbestid and the gbestd in this iteration;

Fig. 1 Delay of the situation when the head part is hindered

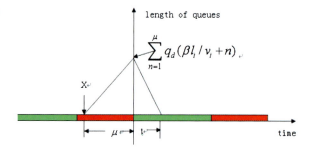

5. Go on to do iterative calculation, update the particle flying speed an position;
6. Finish the circle if the calculation meets the convergence condition, otherwise, go back to ②.

4 Method to Calculate the Delay of Traffic Flow When it Pass the Correlative Intersections

The situation of traffic flow meet the red light when it pass the intersection can be divided into two condition : one is when the head part of the traffic flow reach the Ii intersection, it meet the red light (the first car meet the red light when it reach the stop line), the other is that when the head part reach the Ii, it doesn't meet the red light, and when a part of the flow have already passed the intersection, the light turn red, this defined procession hindered. In the custom researching, when consider the delay of traffic flow when it pass the intersection, they just consider the head part hindered, and the middle part hindered is neglected.

If the traffic flow is hindered, it must at least wait the red light turns green to go across the stop line, with the value of X, we can see the length of red light time which the flow must wait (set as μ). When the flow dissipate, the time that the dissipation costs (set as t), can be obtained from the equations which can be listed out by conservation of the vehicle number (Figs. 1 and 2).

Formula of the basic optimization unit delay:
Delay of the traffic flow when pass the intersection:

$$D_A = 0.5\bigl(S_g + S_r - \beta l_i/v_i + \alpha + t\bigr) \sum_{n=1}^{S_g+S_r-\beta l_i/v_i+\alpha} q_d(\beta l_i/v_i + n) \qquad (7)$$

the delay value of the No. ε unit in one circle is D_ε, with the analysis in previous paper, an conclusion can be given: $D_\varepsilon(\alpha) = D1(\alpha) + D2(\chi)$, phase difference α have a relationship with χ, $\chi = T - \alpha$, T is the traffic signal cycle. So we can get:

$$D_\varepsilon(\alpha) = D1(\alpha) + D2(T - \alpha) \qquad (8)$$

Fig. 2 Delay of the procession hindered situation

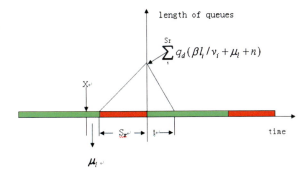

Formula of the total delay of the correlative intersections:

$$D_{system} = \sum_{i=1}^{m}\sum_{\varepsilon=1}^{n-1} D_{i\varepsilon} + \sum_{i=1}^{m}\sum_{j=1}^{2n+2} D_{ij} \qquad (9)$$

Where the first term denotes the sum of the internal traffic flow delay in each subsystem of the correlative intersections, namely the sum of each subsystem delay.

The second term denotes the traffic flow delay which generated when the flow pass the correlative intersections. Formula (9) denotes the total delay of the correlative intersections after m signal cycle times.

5 Application

5.1 Choosing of the Correlative Intersections

The research results of this paper were applied to the correlative intersections in Tianhe district of Guangzhou city, parts of the correlative intersection are : tian he bei —ti yu dong (I_1), tian he bei—tian he dong (I_2), tian he bei—longkouxi (I_3), tian he bei—long kou dong(I_4), tian run—tian he dong(I_5), tian he—ti yu dong(I_6), tian he—tian he dong(I_7). The following Fig. 3 shows this correlative intersections.

5.2 Result and Analysis

The simulation test is set to 10 cycles, totally 2,400 s, the result is (Fig. 4):
Delay of the single intersection obtains through Vissim 3.6 is showed in following tables (Tables 1 and 2):

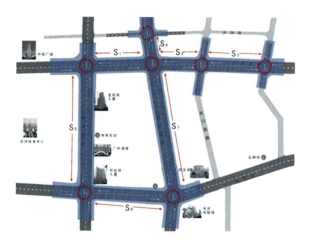

Fig. 3 Subsystem in the correlative intersections

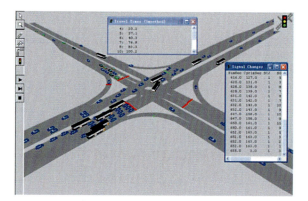

Fig. 4 The simulation test of intersection I_1

Table 1 Result of the simulation test

Intersection code	I_1	I_2	I_3	I_4	I_5	I_6	I_7
Delay of current status (s)	11,231	8,921	531	3,610	951	13,231	13,214
Delay of bidirectional-green-wave numeric method (s)	13,045	6,981	454	2,290	609	8,860	9,765
Delay of our method (s)	8,424	5,226	486	1,945	651	8,901	8,566
Average delay of current status (s/pcu)	26.94	24.77		10.89	22.91	33.50	36.93
Average delay of bidirectional- green-wave numeric method (s/pcu)	31.30	19.38	3.28	6.91	14.67	22.43	27.29
Average delay of our method (s/pcu)	20.21	14.51	3.51	5.87	15.68	22.54	23.94

Note All those results of delay are single cycle result in the simulation test to make the comparison more convenient

Table 2 Delay calculation result of the new model (I)

Control mode	Dynamic correlative intersections Coordinate control					
Subsystem code	S_1	S_2	S_3	S_4	S_5	S_6
Intersection code	$I_1 \leftrightarrow I_2$	$I_2 \leftrightarrow I_3$	$I_2 \leftrightarrow I_5$	$I_3 \leftrightarrow I_4$	$I_1 \leftrightarrow I_6$	$I_2 \leftrightarrow I_7$
Delay of current status (s)	8,504	3,650	272	2,909	9,502	9,998
Delay of bidirectional-green-wave numeric method (s)	5,630	2,172	367	3,505	6,230	7,607
Delay of our method (s)	5,087	1,759	252	1,509	6,025	6,617
Average delay of current status (s/pcu)	42.31	23.55	4.77	13.48	47.98	35.58
Average delay of bidirectional green-wave numeric method (s/pcu)	28.01	14.01	6.44	16.24	31.46	27.07
Average delay of our method (s/pcu)	25.31	11.35	4.42	6.99	28.97	23.55

6 Conclusion

The study results of this paper were applied to the correlative intersections coordinate control in north district of Guangzhou city. Based on simulation experiment and contrast analysis, the results of optimal computation and simulation experiment indicated that the methods proposed in this paper was more effective in reducing delay of vehicles than traditional way and the correctness, effectivity and universality of these models were proved successfully.

References

1. Ceylan H, Bell MGH (2004) Traffic signal timing optimization based on genetic algorithm approach, including driver' routing. Transp Res Part B 38(4):329–342
2. Hounsell NB, McDonald M (2001) Urban network traffic control. Proc Inst Mech Eng Part I 215(14):89–118
3. SCOOT Traffic Handbook (1983) TRRL. pp 45–96
4. Sydney Coordinated Adaptive Traffic System DMR. Australia, pp 36–49
5. FHWA (1983) TRANSYT-7F user's manual. US.DOT. pp. 87–105

Measurement of Electrodermal Activity to Evaluate the Impact of Environmental Complexity on Driver Workload

Maria Seitz, Thomas J. Daun, Andreas Zimmermann and Markus Lienkamp

Abstract In the field of advanced driver assistance systems, researchers and developers make great efforts to reduce drivers' workload and to keep it on an ideal level respectively. In the course of these efforts the physiological detection of workload, also in commercial vehicles, is coming more and more to the fore. A total of 44 driving situations have been identified in an exploratory survey of truck drivers which are relevant for the resulting driver workload. This includes various driving situations as well as driving maneuvers and traffic situations, such as high traffic density or different weather conditions. With regard to objectively measurable parameters, these 44 situations were examined in a study with 37 professional truck drivers in the dynamic truck driving simulator of the Institute of Automotive Technology. In addition to the changes in pupil dilation the electrodermal activity and a subjective assessment based on the Rating Scale of Mental Effort (RSME) were used in order to detect the mental workload. The results were validated in a field trial subsequent to this driving simulator study. At the selection of the test course, particular attention was paid to a high degree of similarities between real-world and simulated roadway sections. The evaluation of the electrodermal activity measured in the real traffic study as an indicator for mental workload revealed the following. The most demanding activity during the performed study was making phone calls, whereby the participants were confronted with a planning task. Paying attention to an accessing car was also a high demanding task with anticipatory requirements with respect to the car driver's behavior. Regarding routine tasks such as crossing intersections, the evaluation of electrodermal activity does not lead to

F2012-I05-003

M. Seitz · T. J. Daun (✉) · M. Lienkamp
Institute of Automotive Technology, Technische Universität München, München, Germany

A. Zimmermann
MAN Truck & Bus AG, Munich, Germany

SAE-China and FISITA (eds.), *Proceedings of the FISITA 2012 World Automotive Congress*, Lecture Notes in Electrical Engineering 200, DOI: 10.1007/978-3-642-33838-0_22, © Springer-Verlag Berlin Heidelberg 2013

statistical differences between the investigated situations. In contrast, a self-report measurement of mental workload conducted in parallel shows sufficient sensitivity and therefore significant differences between the tested maneuvers and environmental conditions. Thus, in order to evaluate everyday driving situations with respect to the mental workload they cause, self-report measures are suggested to be the method of choice. Cognitively high demanding planning or anticipation tasks however can be detected reliably by means of electrodermal activity.

Keywords Driver workload · Driving simulator · Electrodermal activity · Pupil dilation · Rating scale of mental effort

1 Introduction

In recent years, the demands on human information processing have grown steadily with respect to the task of vehicle control. Reasons are the increasing complexity of both the environmental conditions as well as the vehicles per se, which are equipped with more and more driver information and driver assistance systems [1]. Facing this increasing complexity, the driver has to perform reliably in order to reach his destination safely despite the partially increased cognitive demands. One approach aiming at the problem of drivers' cognitive overload is to adapt the behavior of the vehicle systems according to the complexity of the driving situation. In order to assess the complexity of different driving situations, an estimation based on the driver's workload may be used. In this context, the question arises of how demanding different environmental conditions and driving maneuvers are and which measurement methods are suitable to evaluate drivers' mental workload during task fulfillment. The overall goal is thereafter to determine a specific workload index for various situations.

2 Measurement of Mental Workload

The following eight characteristics describe the measurement methods of mental workload and therefore evaluate their suitability for a given application [2, 3]:

- Sensitivity characterizes the ability to express changes in workload.
- Diagnosticity describes to which extend the measurement can be affiliated to a specific source of workload.
- Primary-task intrusion expresses how far the measurement method affects the primary-task performance.
- Implementation requirements is a measure for the effort which is necessary in order to conduct the measurement.

- Operator acceptance characterizes the quality of agreement for a measurement instrument by the subject.
- Selectivity expresses whether a measurement is selective with respect to mental workload and therefore robust against disturbances such as physical load.
- Bandwidth and reliability is the ability to measure the workload in a wide range and independent of the test environment.

Methods in order to measure driver's mental workload can be classified into three groups [2, 3]: Performance measures, subjective or self-report measures, and physiological measures.

Performance measures describe indirect measurement methods of the driver's mental workload. They utilize the fact that the quality in task performance varies with different states of workload. Therefore, workload is reflected by measuring parameters of driver's performance such as lane keeping quality, capability of maintaining a constant speed or clearance, and reaction time. However, since advanced driver assistance systems more and more take over the tasks of lateral and longitudinal control, these indirect parameters do not represent the driver's actual state. Therefore, especially in the future, the measurement of these parameters is not capable of providing appropriate information.

Subjective or self-report measures are generated by questioning the operator using assessment tools and rating scales. According to Muckler and Seven operators are able to assess the state of their workload very well [4]. Self-report measures are easy to implement and very cost efficient. In order to avoid high primary-task intrusion, the questioning of the operator should take place after the task completion. Disadvantages, however, are a variation in interpretation of the assessment questions by different subjects.

Physiological measures intend to detect the effects of mental workload by means of bodily functions (e.g. pupil diameter, cardiologic functions). Opposed to self-report measures, physiological measures enable to observe the workload constantly. Since the measured bodily signals reflect the functions of the central or peripheral nervous system, a deliberate interference by the operator can almost be excluded. A disadvantage of physiological measures is the necessity of special equipment and the risk of impairing the primary-task performance of the subject. Furthermore, it is difficult to clearly attribute the measurements to mental workload.

3 Electrodermal Activity

The term electrodermal activity (EDA) describes changes in the ability of the skin to serve as an electric conductor. A connection between mental processes and changes in the electrical conductance of the skin is known since the late 19th century [5]. Accordingly, with higher mental activity an increased skin conductance is observed. This phenomenon seems to be related to the activity of sweat glands in the skin, which are innervated by the vegetative nervous system. It is

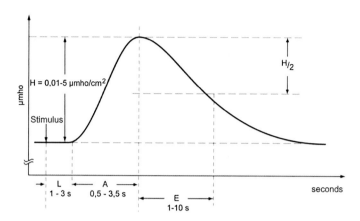

Fig. 1 Skin conductance response [5]; *L* latency; *H* amplitude; *A* rise time; *H/2* half amplitude; *E* half recovery time

assumed that skin conductance increases as a result of a moisturization of the skin triggered by the sweat glands. For the measurement of skin conductance two electrodes are attached on the hand or food, since the effect is strongest at palms and soles, where a high density of sweat glands is found. Direct current in the micro-ampere range and thus not noticeable by subjects is generated to measure the change in voltage and calculate the conductance of the skin in the unit of micro siemens (μS). A distinction is made between tonic and phasic changes of the skin conductance. Tonic electrodermal activities are characterized by stable or slow changes of EDA and reflect the overall psycho-physical activation. Phasic skin conductance responses (SCR) are observed as responses to external stimuli with a latency of 1–3 s. SCR can be seen as event-related momentary increase in skin conductance reflecting the subject's cognitive workload [5]. The SCR curve can be broken down to several components as shown in Fig. 1 [5]. Principally, SCR underlies largely within- and between-subject differences. For detailed information on EDA Bousein is referred to [6].

Measuring skin conductance is a well-established physiological measure with little effort, low-tech and low intrusion, although the process by which mental workload affects the electrodermal activity is not yet completely understood and explored.

4 Approach

In order to measure truck drivers' workload in demanding everyday situations, a three-phase concept was developed (Fig. 2). In the first phase, an exploratory truck driver survey was conducted in order to identify scenarios out of a pool of maneuvers and conditions with high relevance to drivers' mental workload. Since at this phase, the focus was put on qualitative results rather than on quantitative

Measurement of Electrodermal Activity

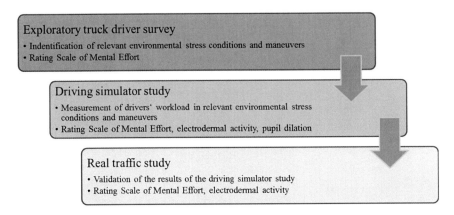

Fig. 2 Three-phase concept to evaluate the impact of environmental complexity on driver workload

results, only self-report measures were collected. Thereafter, phase two served the purpose of generating quantitative results within an experimental driving simulator study utilizing physical measures in addition to self-report measures. In order to ensure an external validity of the simulator study, finally the results have been validated by means of a real-world experiment during the third evaluation phase.

This set of studies thus provides a particularly broad evaluation of journey-related requirements for truck drivers with respect to the mental workload during different driving scenarios. Since multiple measuring methods of mental workload were used during the three phases of evaluation, the results are expected to reveal important findings about the suitability of the single methods in order to generate an index of mental workload at a given driving scenario.

The three phases of evaluation and their study specifications are described hereafter.

4.1 Exploratory Truck Driver Survey

In the first phase, 31 professional truck drivers between 23 and 66 years of age (Ø 43.3, SD 11.33) evaluated 70 situations during an exploratory truck driver survey. The evaluation of a variety of maneuvers like overtaking, parking and crossing of intersections, and conditions such as different weather and traffic conditions were conducted in order to identify stressful driving scenarios of high relevance with respect to drivers' workload. Those 70 situations have been collected previously via surveys, literature research and expert interviews.

An adapted model of the Rating Scale of Mental Effort (RSME) was used in order to measure the participants' self-reported situational workload. The RSME is a standardized uni-dimensional rating scale ranging from 0 to 150. Statements

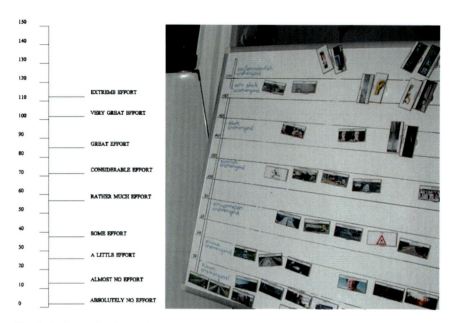

Fig. 3 Rating scale of mental effort, original from Zijlstra (*left*) [7] and adapted (*right*)

related to the subject's invested effort are placed along the scale at several so-called anchor points. Due to the scale design in combination with the indication of the aspect 'invested effort', the RSME is very clear for subjects and thus an appropriate method for self-report measurements [3, 7, 8].

The 70 situations of the exploratory survey were illustrated by pictures in order to ensure the same basis of reference for each subject. Based on this survey, 44 highly relevant scenarios have been identified with respect to driver's workload. The original RSME and the adapted application of the driver survey are shown in Fig. 3 [7, 9]. Details about the methodology and analysis of the exploratory truck driver survey can be found in [9].

4.2 Driving Simulator Study

The exploratory truck driver survey revealed 44 environmental stressful conditions and maneuvers of high relevance. Thereafter, these situations were implemented in the dynamic truck driving simulator of the Institute of Automotive Technology. The driving simulator consists of a hexapod motion platform on which a MAN TGS vehicle cab is mounted. Within the cab, all driver controls are fully operational. The simulator's projection system consists of five projectors covering a horizontal field of view of 210° and three TFT monitors enabling the rear view via wing mirrors (Fig. 4 left).

Measurement of Electrodermal Activity 251

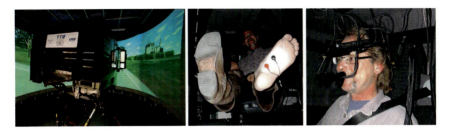

Fig. 4 Truck driving simulator at the Institute of Automotive Technology (*left*), equipment to measure electrodermal activity (*middle*) and pupil dilation (*right*)

During the implementation of the identified situations, particular attention was paid to the isolation of individual factors, which have a specific influence on the workload. The conditions and maneuvers were distributed to a total of seven permutable experimental sections, which are made up of routes on highways in urban and in rural areas. Each of the seven sections contained between 5 and 14 different demanding everyday situations. These situations were placed allowing for a sufficient spatially and temporal clearance between consecutive situations in order to avoid interaction with regard to workload. In this study of repeated measures every truck driver was confronted with all of the seven sections in permuted sequence accounting for a total duration of three hours per experiment. 37 professional male truck drivers between 26 and 69 years old (Ø 47.1; SD 9.8) attended the study [9].

Methods to measure mental workload were once again the RSME and in addition physiological methods measuring electrodermal activity and pupil dilation. Measurement of pupil dilation as an indicator for workload utilizes the fact that the size of human pupils varies with mental stress. A major advantage of this method for determining mental workload is a very low latency of the pupil reaction of 100–200 ms due to a stimulus [10]. Marshall et al. [11] proposes the so-called "Index of Cognitive Activity" (ICA) as metrics to measure mental workload via pupil dilation and is therefore used in the simulator study. This metric is based on the number of strong, rapid and short-term changes in pupil size, which are associated with mental workload and independent from disruptive influences like accommodation or pupillary light reflexes. Schwalm has shown the suitability of the measurement of pupil dilation for an application in a driving simulator [12].

Figure 4 shows the attachment of the measurement equipment to the subject (middle and right). Electrodes to measure electrodermal activity are adhered underneath the left foot. Since the driving simulator is equipped with an automatic transmission no manual clutch engaging is necessary. Thus, this electrode position has a low intrusive impact on the subjects while driving [9].

4.3 Study in Real Traffic

In the third evaluation phase, the results of the driving simulator study were investigated by a study in real traffic in terms of their external validity. Designing a field study consisting of various environmental conditions and driving maneuvers as close as possible to those of the driving simulator study was an extremely challenging task. Due to safety reasons or due to a lack of controllability such as weather or traffic conditions, 10 situations were excluded from the study in real traffic. The resulting 34 situations were distributed to three permutable sections. While one section consisted of highway driving, the other segments were located on rural and urban routes. Each of the latter two sections was interrupted by a resting stop in order to make the experimental design most comparable to the preceding driving simulator study. In a within-subject design, a subset of the subjects of the driving simulator study completed real-world study. The within-subject design of this trial offers the advantage of a higher statistical strength while remaining the sample size. 12 drivers between 29 and 69 years old (Ø 44.9; SD 9.8) covered a distance of 121 km at the study in real traffic, which took about 5 h in total [13].

During the study, a truck of the type MAN TGX 18.440 served as the experimental vehicle pulling a trailer and thus allowed for a comparable set-up with respect to the driving simulator study (e.g. automatic transmission, empty trailer). Measurement methods to evaluate driver workload were in this case the self-report measure RSME as well as the physiological measure of electrodermal activity. Pupil dilation was not recorded due to the complex measurement equipment and its impact on transport safety. Again, the electrodes for the measurement of electrodermal activity were attached at the truck driver's left foot. The experiment vehicle and the measurement set-up are shown in Fig. 5 [13].

5 Data Evaluation and Results of The Study in Real Traffic

In order to evaluate the SCR data the following approach was used. During the experiment, the SCR data was logged continuously once per second in which the corresponding situations have been marked using a trigger. At the evaluation process the amplitude of the marked situations was investigated. The situational timeframe under investigation was extended by a latency buffer of 3 s. Excluded from the data evaluation were situations which could not be performed as planned (e.g. unintended interferences by other road users during the real traffic study), contained measurement errors (e.g. detached electrodes) or did not trigger an appropriate physiological reaction of at least 2 μS. These restrictions led to the exclusion of 8 situations from data evaluation. Since the SCR data is subject to inter-individual differences, the amplitude level has to be normalized. For this purpose the situational amplitude levels of a single subject were normalized using

Fig. 5 Experimental vehicle (*left*) and equipment to measure electrodermal activity in the real traffic study (*right*)

the subject's maximum amplitude level of all situations. A workload index is made up by the median of all valid SCR values of a specific situation. The situations were classified into three groups: highway situations, country road situations, and urban situations. Comparisons between situations regarding workload are just drawn within a distinct group in order to keep the complexity limited. Whether the differences between situations of the real-world study are significant was evaluated using the Wilcoxon-signed-rank test for matched pairs.

The differences for seven country road situations as well as 14 urban situations do not provide statistical evidence for the majority of these situations. Table 1 provides an overview of the statistical results of the situations, which took place during highway driving.

The situation "making phone calls" shows significant differences to all other situations except "paying attention to an accessing car" on the acceleration lane. In the latter situation the driver of the accessing car did not indicate clearly whether he intends to enter the highway in front of or behind the truck. In turn the Wilcoxon-signed-rank test confirms significant differences to the situations "accessing the highway" and "simply driving on the highway" respectively a tendency towards significance for the situation "searching for a parking space".

The highest workload is induced while making a phone call (Fig. 6). During the phone call the participants were confronted with a planning task. They had to provide information whether they are able to accept a transport assignment at an upcoming day and estimate the time needed in order to fulfill the assignment. Paying attention to an accessing car requires effort in anticipation of the car driver's behavior and thus causes a high mental workload.

Both of the previous mentioned situations require planning activity and therefore lead to a high mental workload which also is reflected in the physiological measurement data. The measurements of electrodermal activity of all other situations, including urban and country road turning situations, do not provide sufficient sensitivity and thus do not lead to statistically sound differences in SCR. These situations are solely frequently executed driving tasks, for which a high degree of training routine can be assumed.

Table 1 Results of the Wilcoxon-signed-rank test (z, p) for highway situations of the real traffic study

	Making phone calls	Paying attention to an accessing car	Air condition adjustment	Accessing the highway	Searching for a parking space	Simply driving on the highway
Making phone calls		−0.105	*−2.24	*−2.521	*−2.521	*−2.366
Paying attention to an accessing car			−1.572	*2.201	~−1.782	*−2.023
Air condition adjustment				−1.12	−1.352	~−1.859
Accessing the highway					−0.507	~−1.690
Searching for a parking space						−0.734

*$p < 0.05$
**$p < 0.01$
~$p < 0.1$

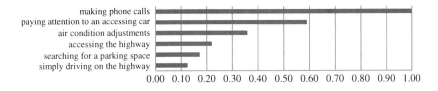

Fig. 6 Median of SCR values for situations while driving on highway

6 Discussion and Conclusion

Based on the results of the real traffic study, it can be assumed that the measurements of electrodermal activity do not provide sufficient sensitivity for workload during routine tasks. In contrast the RSME data shows to be highly sensitive for the majority of all evaluated situations. This corresponds to the results of De Waard [3] who also attests self-report measurements to a higher general sensitivity. Therefore in order to create a comprehensive evaluation of environmental stressful conditions and maneuvers by means of a generalized workload index self-report measures such as RSME are suggested to be the method of choice. However, electrodermal activity is suitable to detect cognitively high demanding tasks such as planning and anticipating.

References

1. Bundesamt für Güterverkehr. Marktbeobachtung Güterverkehr. Auswertung der Arbeitsbedingungen in Güterverkehr und Logistik 2011-I, 2012
2. O'Donnell RD, Eggemeier FT (1986) Workload assessment methodology. In: Boff KR, Kaufman L, Thomas JP (eds) Handbook of perception and human performance, Volume II, cognitive processes and performance: 42/1–42/49
3. De Waard D (1996) The measurement of drivers' mental workload. University of Groningen, Haren
4. Muckler FA, Seven SA (1992) Selecting performance measures: 'objective' versus 'subjective' measurement. Hum Factor 34:441–455
5. Gramann K, Schandry R (2009) Psychophysiologie. Körperliche Indikatoren psychischen Geschehens. Beltz Verlag, Weinheim
6. Boucsein W (2011) Electrodermal activity. Springer, New York
7. Zijlstra FRH (1993) Efficiency in work behavior. A design approach for modern tools. Delft University Press, Delft
8. Eilers K, Nachreiner F, Hänecke K (1986) Entwicklung und Überprüfung einer Skala zur Erfassung subjektiv erlebter Anstrengung. Zeitschrift für Arbeitswissenschaft. 40:215–224

9. Seitz M, Zimmermann A et al (2011) Pupillometrie als Werkzeug zur Erfassung der Fahrerbeanspruchung im Nutzfahrzeug. Der Fahrer im 21. Jahrhundert, Braunschweig 2011
10. Beatty J (1982) Task-evoked pupillary responses, processing load, and the structure of processing resources. Psychol Bull 91(2):S.276–292
11. Marshall SP (2002) The index of cognitive activity: measuring cognitive workload. In: 7th IEEE conference on human factors and power plants, New York, 2002
12. Schwalm M (2009) Pupillometrie als Methode zur Erfassung mentaler Beanspruchungen im automotiven Kontext. Universität des Saarlandes, Saarbrücken
13. Seitz M, Feldmeier D, Zimmermann A et al (2012) Eine Feldstudie zur Erfassung der kognitiven Beanspruchung von Lkw-Fahrern. Commercial Vehicle Technology Symposium, Kaiserslautern, 2012

Part VI
Other

Study on Engine Idle Speed Control Considering Vehicle Power Balance

Feng Gao, Qiang Zhang, Daquan Zhang and Wen He

Abstract By analyzing the characteristics of a battery discharge, a new power balance automatic control strategy at idle condition was designed. The feedback signal of electric power load is cancelled and the battery state is estimated only by voltage. Furthermore, application problems, such as integration with the normal engine control logic, etc. were studied. By integrating into ECU, the idle power balance automatic control function can be realized without adding any hardware cost. Vehicle test results show that the control logic can modify idle speed automatically and stably according to electrical load power.

Keywords Power balance · Engine control · Idle speed control · Electrical load estimation · State of charge

1 Introduction

Normally the lower the idle speed is, the performance of fuel consumption, emission and Noise Vibration and Harshness (NVH) is better. But to ensure the power balance and the stable of engine, working at higher idle speed is better. The technology, which modifies the idle speed automatically according to the power load, can improve engine performances. And it is already widely used in middle/high class cars.

F2012-I06-002

F. Gao (✉) · Q. Zhang · D. Zhang · W. He
Automobile Engineering Institute of Changan Automobile Ltd, Chongqing, P.R. China

The widely used control method is described as follows [1, 2]. When the electrical power load is small, idle speed is set to a lower one, otherwise, using a higher idle speed. Always the possible desired idle speed is set to several levels, which are calibrated previously. This control method is simple, but some special equipment is needed to measure the electrical power load. One choice is a battery sensor, which can measure voltage, current and temperature of battery. Another one is a special generator, which can feedback the electrical power load directly.

In China, more than 2.5 million mini cars are produced per year. It is very important to reduce their fuel consumption and emission by power management technology. But the consumers can hardly afford the added costs. To avoid the conflict between costs and performances, a new idle speed control method is designed in this paper. It can estimate the electrical power load only by the battery voltage. No extra hardware is needed and the software, which estimates power load and controls the idle speed, can be integrated into engine control unit. The proposed method has been verified by a prototype mini car. The results show that the proposed method can control the engine idle speed according to electrical power load automatically. The discharge of battery is controlled in an acceptable range and the transition process is smooth.

2 Idle Speed Control Method

2.1 Electrical Power Load Calculation

The voltage of a battery satisfies the following equation [3–5].

$$V_t = V_e - I \times R_b. \tag{1}$$

Where V_e is the electromotive force, V_t is the open circuit voltage, I is the charge/discharge current and R_b is the resistance. V_e and R_b are functions of State of Charge (SOC) of battery. The objective of the idle power balance control is to ensure that the battery will not be over discharged, which means that the SOC of the battery varies small. So the V_e and R_b can be considered as constant and the power consumption of the battery, Q, can be estimated by

$$Q = \frac{1}{R_b} \int (V_t - V_e) dt. \tag{2}$$

According to the above analysis, the following power balance control algorithm at idle condition for power management system is designed and is shown in Fig. 1.

The reference voltage is calibrated. The acceptable discharge range of battery is larger and the reference voltage is lower. The battery is considered charge when the voltage of battery is greater than the reference value. Considering accumulation error and its inaccuracy, the following restrictions are used:

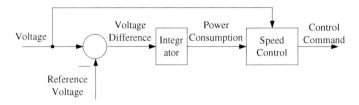

Fig. 1 Engine idle speed control system

(1) The upper limit of the integrator is set to zero.
   In theory the discharge amount can not be greater than charge amount, which means the integration of the battery current is always smaller than zero.
(2) The integrator has a lower limit, which is calibrated according to the allowable discharge amount of battery.
   The lower limit is smaller, the acceptable discharge amount of battery is bigger.

2.2 Speed Increase/Decrease Determination

Since the power consumption calculated by voltage difference is not precise, it is improper to use this value to determine increase/decrease of idle speed directly. Both the power consumption and the battery voltage are take into count. The idle speed increase/decrease logic is:

(1) When the power consumption is smaller than the integrator lower limit, the control command is to increase idle speed,
(2) When the power consumption is greater than zero and the battery voltage is greater than the threshold, which is calibrated, the control command is to decrease idle speed,
(3) Otherwise, the control command is to hold the idle speed.

In the second rule, the voltage threshold is the battery voltage in the condition that the battery is full and charging a little, which is between 14–14.5 V usually. This added condition is to ensure that the idle speed only decreases when the battery is full.

In this paper, the integration of voltage difference is used to determine the electrical power load, not the voltage directly. It has the following advantages:

(1) Avoiding the frequently changing of idle speed, which may occur when electrical loads turn on or off.
(2) Comparing with battery voltage, the integration of voltage difference is more closely to the real power consumption.

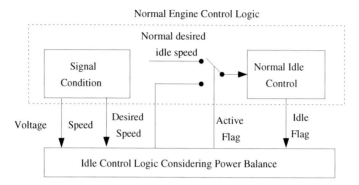

Fig. 2 Integration with normal engine control logic

3 Practical Application Problems

In the above section, the main idle speed control method is introduced. But in practical application some other problems should be tackled, such as the smoothness of the transition process, integration with the engine control logic, active/inactive logic of the idle speed control.

3.1 State of the control method

The proposed method is to solve the electrical power balance problem at engine idle condition, so it is only active when the engine is warm and in idle condition. Considering determination of the initial idle speed when the logic is activated, the power consumption calculation function should be run when engine is in other condition.

When the engine is in idle state, the initial idle speed can be determined from the calculated power consumption. Normally, the initially idle speed is linear with the power consumption. The absolute value of the power consumption is bigger, the added idled speed is bigger.

3.2 Integrate with Engine Control Logic

The interaction between the proposed idle control logi and the normal engine control logic is showed in Fig. 2.

The normal idle control logic of engine control unit outputs idle flag, which is true when engine is in warm and idle condition [6, 7]. And the idle control logic designed in this paper is active or inactive according to:

(1) When the idle flag is true, the designed idle control function runs. Its active flag is set and the engine idle speed is controlled by the desired speed output by the proposed idle control function.
(2) Otherwise the active flag is reset and the engine idle speed is controlled by the normal engine control logic. At this condition, the power consumption calculation function runs normally.

Furthermore, the normal engine control logic outputs engine speed, which is used to calculated the new desired idle speed. Both the desired engine speed should be saturated by the normal idle control logic. The smoothness of the transition process should be considered when calculating desired idle speed, which is discussed in the following section.

3.3 Transition Process

The idle speed adjusting process should be smooth. This affects NVH greatly. The idle speed is adjusted periodically and the logic is:

(1) When the control command is speed increase, the desired idle speed is derived by adding the engine speed with a constant.
(2) When the control command is speed decrease, the desired idle speed is derived by subtracting a constant from the engine speed.
(3) Otherwise the desired idle speed is equal to the engine speed.

The adjusting period and the constant are calibration values. The shorter the period is, the bigger the constant is, then the adjusting speed of the idle speed is quicker, the risk of getting bad NVH performances is much more. In this paper, the new desired idle speed is calculated by adding/subtracting a constant value from the actual engine speed and not by adding the adjusting value to the normal desired idle speed directly. It can avoid the desired idle speed running outside of the designed region far, which may result in sudden changes in engine speed.

4 Vehicle Experiments

The above sections introduce the proposed idle control logic, which can confirm the power balance in idle condition. In this section it is verified by vehicle tests.

4.1 Prototype System

The prototype system of the idle control system for vehicle test is shown in Fig. 3.

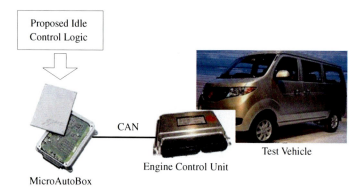

Fig. 3 Prototype system for vehicle test

The proposed idle speed control logic is modeled by Matlab/Simulink and runs on MicroAutoBox [8]. The signals between the proposed logic and the normal engine control logic are transmitted by CAN. The prototype system is equipped in a mini car and the calibrated values are set as:

(1) The reference voltage is 12.8 V, which is derived by the relationship between the SOC and the open voltage.
(2) The lower limit of the integrator, which calculates the power consumption, is −10.
(3) The voltage threshold of the second rules is 14.2 V. It is derived by the nominal output voltage of generator.
(4) The period of adjusting the idle speed is 20 s and the adjusting value is 10 r/m.
(5) The upper and lower limit of the idle speed control region is 1,000 r/m and 750 r/m respectively.

4.2 Experiment Results

Vehicle test conditions include:

(1) Transition between active and inactive modes.
(2) Idle speed increase and decrease process.
(3) Big and small discharge current.
(4) Turn on/off power load frequently.

The typical test curves are shown in Fig. 4.

Figure 4a is the charging/discharging current of battery. Positive value means charge and negative means discharge. Figure 4c is the proposed idle speed control logic active flag. One is active and zero is inactive. 0–80 s is the start process of

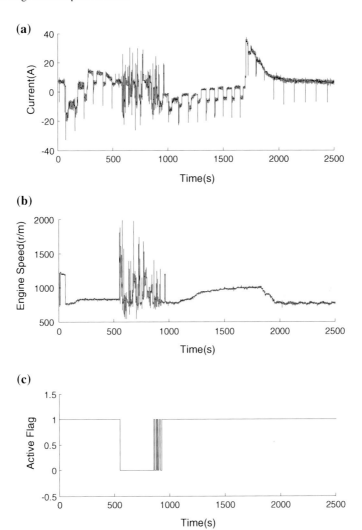

Fig. 4 Vehicle test results (**a**) Charging/Discharging current of battery (**b**) Engine Speed (**c**) Idle speed control logic active flag

engine and the idle control function is inactive. From 80 to 200 s, a load, whose power is big, is turned on and engine idle speed is increased from 750 r/m to 800 r/m immediately. Then the battery is in charging state and the idle speed stays at 800 r/m until the battery is full. From 550 to 950 s, the proposed control logic changes between active and inactive because of the engine state changes. When the proposed control logic is inactive, engine is controlled by the normal control logic. After 950 s a load, whose power is small, is turned on. The battery discharges about 150 s continuously and then the idle speed starts increasing smoothly until reaches the upper limit. From 1,750 s the electrical load is turned off and the battery is charging.

From 1,800 s the idle speed begins decreasing smoothly until the lower limit. During the tests, some loads are turned on/off frequently and there is no fluctuating of engine speed.

5 Conclusions

To reduce the engine idle speed as much as possible while confirming the power balance, an idle speed control method only by voltage is designed based on the already available electric component in vehicle. The effectiveness of the proposed method has been verified by vehicle tests and the results show:

(1) The proposed method can estimate the power consumption of the battery. The engine speed can be adjusted automatically according to the electrical power load and the transition process is smooth.
(2) The proposed control method can be integrated with normal engine control unit and all the control functions work correctly. The new added control function has no negative impact on engine performances.
(3) By applying this control method, the idle speed control function considering power balance can be realized without adding any hardware.

References

1. Daniel JM, Nad ASD (2006) An method by increasing engine idle speed to allow battery charging. China, CN101008356A
2. Mehrdad E, Yimin Gao, Ali E (2010) Modern electric hybrid electric and fuel cell vehicles-fundamentals theory and design. CRC Press, London
3. BOSCH (2004) Automotive electric and electronic of BOSCH. Beijing Institute of Technology Publishing House, Beijing
4. Ribbens WB (1998) Understanding automotive electronics. Butterworth-Heinemann, Woburn
5. Zhang Q, Gao F, Hou J (2009) Simulation study of charging balance for vehicle electric power system. Society of Automotive Engineering, Beijing, 2009, pp 463–467
6. Zhang X (2004) Vehicle engine control technology. Industrial Publishing House, Beijing
7. Guzzella Lino, Onder CH (2007) Introduction to modeling and control of IC engine system. Springer Press, New York
8. Li G-Y (2007) Intelligent control and application of matlab in engine control. Electrical Publishing House, Beijing

Intelligent Functionalities for Fully Electric Vehicles

Adrian Zlocki, Qihui Huang, Mimoun Ghaouty El, Lutz Eckstein and Holmer-Geert Grundmann

Abstract "Intelligent Dynamics for fully Electric Vehicles" (ID4EV) is an European research project within the 7th framework programme co-funded by the European Commission. The objective of the ID4EV project is to develop energy efficient and safe braking and chassis systems as well as intelligent functionalities and new Human Machine Interface (HMI) concepts for the needs of Fully Electric Vehicles (FEVs), in order to improve active safety and comfort for a faster introduction of FEVs. In this paper the range calculation and the HMI concepts including a range problem solver for FEVs of the ID4EV project is introduced and described. First implantation results are shown.

Keywords Fully electric vehicles · Range calculation · HMI concepts · Advanced driver assistance systems · Intelligent functionalities

Fully Electric Vehicles (FEV) will offer a high potential for the long-term reduction of CO_2-emissions. Current development efforts in FEV are mainly focused on cost and weight reduction of electric drive trains and energy storage systems as well as on extending the operating range. However, customer acceptance regarding safety and comfort has to be taken into account as well, in order to accelerate the introduction of FEVs.

"Intelligent Dynamics for fully Electric Vehicles" (ID4EV) is an European research project within the 7th framework programme co-funded by the European

---

F2012-I06-005

A. Zlocki (✉) · Q. Huang · M. G. El · L. Eckstein
Institut für Kraftfahrzeuge RWTH Aachen University, Aachen, Germany

H.-G. Grundmann
Continental Automotive GmbH, Regensburg, Germany

Commission [1]. The objective of the ID4EV project is to develop energy efficient and safe braking and chassis systems as well as intelligent functionalities and new Human Machine Interface (HMI) concepts for the needs of FEV, in order to improve active safety and comfort for a faster introduction of Fully Electric Vehicles [2].

The work package "Intelligent Functionality by Cooperative Interaction for fully Electric vehicles (IF4EV)" aims to provide intelligent functionalities to the driver, which increase the usability and attractiveness of FEVs. The driver can choose from a set of given profiles reflecting his preference for the actual trip (energy efficient, comfortable etc.). Based on the driver's choice a Profile Manager configures all components of the vehicle in an optimal way in order to realize the chosen profile. An advanced HMI, which is adapted to the EV architecture, is developed to meet the needs of the driver in a FEV.

For electric vehicles it is essential to know as accurate as possible, what the remaining driving range is. In IF4EV a model-based estimation for the remaining range is implemented in order to achieve high accuracy. For this purpose a vehicle longitudinal model is developed to predict energy consumption under specific conditions. Digital map data is used to allow integration of road data, e.g. road topology and traffic rules, in range calculation. Besides this, driving behaviour (e.g. driven speeds, accelerations etc.) of each particular driver is estimated based on historic data. In case the predicted driving range is not sufficient to reach the next charging station or destination, support or advice could be provided to the driver in order to save enough energy by e.g. turning off some auxiliary consumers to reach the desired destination.

1 Range Calculation Based on a Vehicle Model and a Digital Map

In work package "IF4EV" two navigation cores are used for the realization of intelligent navigation functionalities as shown Fig. 1. The foreground core is used to provide navigation instructions to the driver. The background core is used to provide route events (e.g. speed limit, inclination etc.) to the range calculation algorithm to enable predictive range estimation, and to monitor the distance to the final destination as well as alternative destinations (e.g. charging spots, home position etc.). In case that the predicted driving range is not sufficient to reach the next charging station or destination, support or advice can be provided to the driver in order to save enough energy by e.g. turning off some auxiliary consumers to reach the desired destination. The activities of the two navigation cores are coordinated by the so called "Navigation Control" algorithm.

A model-based estimation for the remaining range is implemented in order to achieve high accuracy range calculation. The vehicle longitudinal dynamics is modelled in an Matlab/Simulink environment (see Fig. 2). Furthermore essential

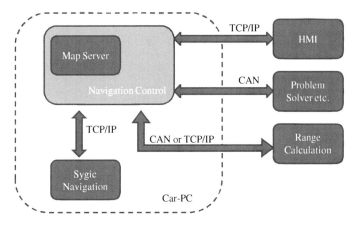

Fig. 1 Navigation control software architecture

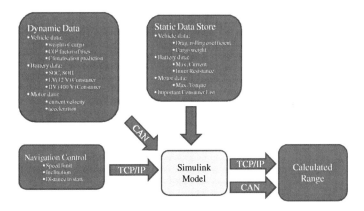

Fig. 2 Overview of input parameters for the range calculation model

components of the drive train: transmission, battery and electric motor are modelled as well. Inputs for these models are dynamic data and static data. In order to consider the road information, e.g. speed limit or slope, which have significant influence on the range, in the range calculation, an ADAS digital map is used. This is integrated in the "Navigation Control" software, which communicates with the Range Calculation via UPD and provides the required information depending on the selected profile by driver.

Simulation results for the Range Calculation model of a test scenario are shown in Fig. 3 as an example. The driven route includes different speed limits and inclinations, as shown in the first two figures. The initial SOC of the battery is 90 %. The third figure shows the course of the SOC on the route. After a distance of about 30 km up to 50 km there is a declination (inclination value up to -10 %). In this area the SOC value increases due to recuperation of the battery. The simulation results verify the functionality of the Range Calculation modelling.

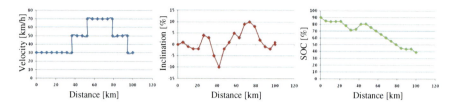

Fig. 3 Results for range calculation taking charging sports into account

In order to apply the developed models in a vehicle by a driver several profiles are developed to meet the driver's needs. These are:

1. City
   The profile "City" can be used, if the user wants to drive in the city or area around his home and a maximum of drive comfort is required. All systems will be configured to allow the highest possible comfort to the user. In addition the system monitors the remaining range continuously to ensure, that the home or defined alternative destination will be reached without an additional charging stop.
2. Travel
   The profile "Travel" can be used, if the user wants to drive to a defined destination and a maximum driving range is required. All systems will be configured to provide the most energy efficient functionality. In addition the system monitors the remaining range continuously to inform the driver correctly if the defined destination will be reached with or without additional charging stops.
3. Fun
   The profile "Fun" can be used, if the user just wants to drive for fun or without the need of maximum driving range. The system provides three alternatives of navigation support "Back to home", "Charging stations along the road", "Defined destination". All systems will be configured to allow the highest driving performance to the user, without sacrificing the safety of the vehicle. In addition the system supervises the range vs. the distance to home, charging spot or destination continuously to ensure, that the wished destination will be reached.
4. Commuter
   The profile "Commuter" will be activated by the system automatically, if a regular recurrent trip is detected out of GPS position, user, destination, time and time of day. In addition the user can provide the system information about his commuter behaviour, so that the detection gets more reliable. All systems will be configured to allow the highest possible driving comfort to the user and requirements of the trip. In addition the system supervises the driven route continuously to detect deviation from the planned commuter route, and check the remaining range versus the know commuter destination to ensure, that the commuter destination will be reached.

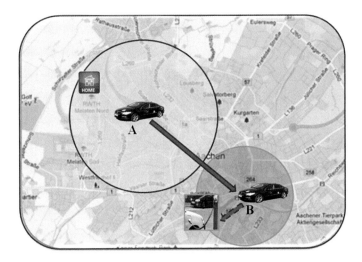

Fig. 4 Example scenario for profile "City"

5. Limp Home
   The profile "Limp Home" will be activated by the Problem Solver algorithm, if no solution under normal driving conditions is available. By this system profile all systems will be configured to the minimum power consumption, which make it possible to reach the next possibility to charge the EV can be reached. The feature has to be activated by the user intentionally. This profile can only be stopped by recharging the EV.
6. MyProfile
   The profile "MyProfile" allows the user to configure his/her own profile out of the configuration options of the profiles "Travel", "City" and "Fun". This use case allows the user to adopt his/her EV best to his/her special purpose of use and increases the attractiveness of the EV.

In different profiles different information are extracted from the digital map by the Navigation Control. The user scenarios "City" and "Travel" are introduced as examples in the following.

If the user drives in the city and activates the "City" profile, the system will monitor the available range and the distance to the home location set by the driver and warns the driver, if the SOC of the battery is high enough to return home, as shown in Fig. 4 at point A. If the user ignores the warning and drives further (point B in the figure), the available energy will not allow the user to return home. The system will then search for the next available charging spot and provides navigation instruction to guide the user to find the charging spot.

In the "Travel" profile, the user can input a destination, which might be located out of the current available range of the vehicle. In this case the system will plan the required intermediate stop(s) alone the planned route, where the user can charge the vehicle. For each stop, several possible charging spot (if available) will

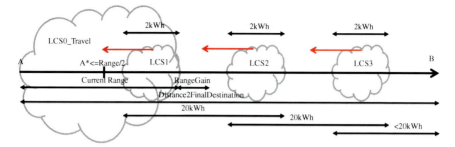

Fig. 5 Generation of charging spot clouds along the planed route in the "Travel" profile

be provided for the user's choice. By doing this, the user has an overview already at the beginning of the trip. Figure 5 shows an example for the generation of the Charging Spot clouds along a planned route with of necessary power values given as examples.

2 HMI Concept

The advanced HMI concept developed in IF4EV is based on the set of defined profiles mentioned above. These profiles support the driver of a FEV in his daily use of the vehicle. Figure 6 shows the HMI for the profiles "Travel", "City" and "Fun" as examples.

In the different profiles on one hand the skinning of the HMI will be adjusted to the driving purpose and on the other hand the system performance will be adjusted to long range, maximum comfort or maximum driving performance. In addition the navigation behaviour will we configured to make sure, that the system always has a defined destination. This is important for a special feature, which is called "Range Problem Solver". This feature compares the distance to destination with the current range and generates solutions, which increase the range of the vehicle by changing the configuration of the system. These solutions are presented by the HMI and the driver can accept or reject the provided solution. In addition the driver gets information about the main power consumer in the system, so that he/she can decide to switch of consumer to increase the range (see Fig. 7).

For the "Range Problem Solver" the predictive range calculation has to provide information about the range, which is linked to each power consumer. This information will be used to generate, depending from the selected profile, the best solution to increase the range. If the configuration of the power consumer is not enough to reach the destination, the HMI provides information about charging spots in reach (see Fig. 8).

Intelligent Functionalities for Fully Electric Vehicles 273

Fig. 6 Travel, city and fun main menu screen

Fig. 7 Current power consumption screen

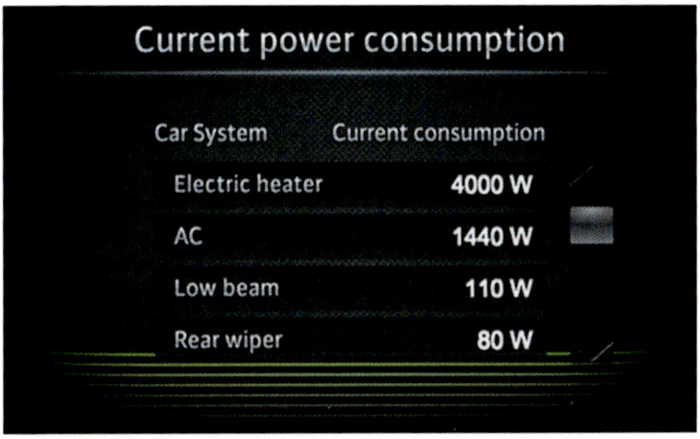

Fig. 8 Solution screen

3 Conclusion

In the frame of the European research project ID4EV a Range Calculation model for fully electric vehicles was developed. The developed system is based on information of vehicle components and digital map data. The interface to the driver is realised in terms of several driving profiles which reflect different driving characteristics (trip length, power usage etc.). By means of an advanced HMI concept useful information is provided to the drivers of FEVs. This increases the comfort and thus the usability and attractiveness of FEVs.

The Range Calculation and the HMI are integrated in a prototype vehicle. An assessment of the system with test persons will also be conducted during the last stage of the project. Objective opinions of different drivers towards the system will be collected for the evaluation and improvement of the system.

Acknowledgments The research leading to these results has received funding from the European Community's Seventh Framework Programme (FP7/2007-2013) under grant agreement n° 260070.

Project consortium: Continental, Forschungsgesellschaft Kraftfahrwesen mbh Aachen (fka), RENAULT, ZF Friedrichshafen AG, IDIADA, CHALMES, TNO, ICOOR.

References

1. Sandro B (2011) ID4EV—intelligent dynamics for fully electric vehicles Barcelona motor show. Barcelona, May 18, 2011
2. ID4EV project website—http://www.id4ev.eu

Motion Stabilizing Controller of Off-Road Unmanned Wheel Vehicle in 3 Dimensional Space

Yue Ma, Changle Xiang, Qingdong Yan and Quanmin Zhu

Abstract Stabilization of unmanned ground vehicle (UGV) in three dimensional space is of exceeding importance. In this paper, stabilizing controller was presented. To reveal the behaviour of UGV, models of major modules of UGV and the mechanism of disturbances applied on were discussed. Subsequently, PID method was employed to compensate the impacts of disturbances and simulation results proved the validity for disturbance incited by slope force.

Keywords Off-road · Unmanned wheel vehicle · Stabilizing control · 3-dimensional space

1 Introduction

The unmanned vehicle become more and more important in present world, which have acted as explorers, rescuers, detectors, servants [1], and tracked unmanned ground vehicle (UGV) has plenty of prominent advantages in the practical area, lots of subject of research focus on it.

As a UGV was expected to work in self-organized manner, consequently, knowing where it is, where the destination is, how it can get there and keeping

---

F2012-106-006

Y. Ma (✉) · C. Xiang · Q. Yan
Beijing Institue of Technology, Beijing, China

Q. Zhu
The University of the West of England, Bristol, UK

stability when it is running are of importance for a UGV. Furthermore, when a UGV works outdoors and has to be subject to space constraints in 6-DOFs, in which environment, lots of disturbances make the objects mentioned above difficult to realize. Disturbances in three dimensional space could be summarized as follow:

(1) Varied adherence coefficient of varied type of ground,
(2) Varied resisting force on magnitude and direction coming from varied terrain,
(3) Random disturbance under uncertainty environment.

Under the effects of disturbances, UGV is prone to deviate from predefined routine, which would lead UGV into uncertainty and dangerous situation in all probability. That is, stability of motion trajectory of UGV is lost.

With regard to theoretical level, general motion stability theory [2] elaborate on the rules which a dynamic system should respect. As it is a very important property for most of dynamic systems, theories of stability of motion were applied broadly, for instance, in general plants or mechanical systems to keep the system stabilize to equilibrium working point [3–5].

Automobile industry paid much attention to the handling stability when car is running in curve path with high speed [6, 7]. Based on the analysis of stability, driving assistant system, such as electronic stability program (ESP) was developed rapidly and became the standard equipment of. Without ESP, when car is running with left wheel on icy road and right wheel on blacktop, side slip would drive the car overturn. While equipped with ESP, under the same condition above, ESP would re-allocate driving force to left and right wheel to keep car running stable.

Compared with theoretical definition of stability, the motion stability of UGV can be summarised as the error between the motion states and pre-defined motion states have to be converged to zero with time goes on.

In this paper, dynamical model of UGV was established. And, to reveal the mechanism of disturbances applied on UGV, two kinds of representative disturbances (slope and general disturbances in yaw motion) will be discussed in depth, which will be the fundamental objects to be resolved. Subsequently, a PID based controller will be employed to compensate the impacts of disturbances.

2 Modelling of UGV

Unexceptionally, modelling is the first key procedure in research of UGV, which is the foundation for the subsequent research, i.e. PID controller design.

2.1 Scheme of UGV

Generally speaking, a UGV is a class of complex systems consisting of lots of mechanical, electric and electronic parts. Fig. 1 present a scheme of UGV

Fig. 1 Prototype of UGV

designed by Beijing Institute of Technology, which is involved the following modules according the function: (1) energy source module, (2) control module, (3) power and driving module, (4) main body and (5) auxiliary module.

In this paper, models of driving module and main body are built up, and others are neglected because they contribute tiny affection on the dynamic behaviour of UGV in short period.

2.2 Modelling of Driving Modules

Driving modules are very important for UGV, the kernel function of which is to convert energy from electronic energy into mechanical energy. There are several kinds of driving equipment available for UGV, diesel motor, DC motor, permanent magnitude synchronous motor (PMSM) motor and brushless DC (BLDC) motor. Survey on the characteristics of motors indicates that many advantages promote the application of BLDC motor in UGV. Therefore, it is chosen as the main power of the UGV in this research project.

According to [8], BLDC motor could be denoted as:

$$\begin{bmatrix} u_a \\ u_b \\ u_c \end{bmatrix} = \begin{bmatrix} r & 0 & 0 \\ 0 & r & 0 \\ 0 & 0 & r \end{bmatrix} \begin{bmatrix} i_a \\ i_b \\ i_c \end{bmatrix} + \begin{bmatrix} L-M & 0 & 0 \\ 0 & L-M & 0 \\ 0 & 0 & L-M \end{bmatrix} \begin{bmatrix} \dot{i}_a \\ \dot{i}_b \\ \dot{i}_c \end{bmatrix} + \begin{bmatrix} e_a \\ e_b \\ e_c \end{bmatrix} \quad (1)$$

where u_a, u_b, and u_c are phase voltages of every winding, i_a, i_b, and i_c are phase currents of every winding, L is coefficient of self-inductance, r_a, r_b, and r_c is the internal resistance of one winding, M is coefficient of mutual-inductance, and finally, e_a, e_b and e_c are back-electromotive force (EMF) of every winding, which are functions of angular velocity of the rotor shaft, accordingly:

$$e_i = K_e \omega_m \qquad i = a, b, c \tag{2}$$

where K_e is named back-EMF constant.

And the driving torque of BLDC motor is

$$T_m = \frac{1}{\omega_m}(e_a i_a + e_b i_b + e_c i_c) \tag{3}$$

in which ω_m stands for the angular velocity of motor shaft.

Finally, both driving torques of BLDC motors will be converted into driving forces by gearbox, sprockets and tracks, which is presented as:

$$\begin{aligned} F_{t1} &= T_{m1}/(\eta_b i_b r_z) \\ &\text{and} \\ F_{t1} &= T_{m2}/(\eta_b i_b r_z) \end{aligned} \tag{4}$$

in which,

η_b: total efficiency of gearbox, sprockets and tracks;
i_b: total transmission ratio of gearbox, sprockets and tracks;
r_z: radius of sprocket.

2.3 Modelling of Main Body

Compared with most of proposed research in robotics and vehicle industry, which are carried out in 2-D space, body model of UGV is described in 3-D space in this paper. And in this condition, motions in six-DOFs should be considered. According to Newton–Euler formulation, general motion of UGV could be summarized as

$$\frac{d}{dt}\mathbf{H}_B = \mathbf{M}_B + m(\omega \times \mathbf{r}_{G/B}) \times \mathbf{r}_B \tag{5}$$

In which
$\mathbf{H}_B$: momentum of rigid body about arbitrary base point B.
$\mathbf{M}_B$: sum of moments about arbitrary base point B.
$\mathbf{r}_{G/B}$: distance vector of from CG to point B.
ω: angular velocity of rigid body.
$\mathbf{v}_G$: translational velocity of CG.
$\mathbf{v}_B$: relative velocity of acting point of forces to CG.

Substituted with corresponding parameters of the UGV designed in my lab, its dynamic equations should be:

$$F_{t1}\mathbf{i} + F_{t2}\mathbf{i} - F_f\mathbf{i} - mg\sin\theta\mathbf{i}$$
$$= \delta m\left(\frac{dv_x}{dt} + v_z\omega_y - v_y\omega_z\right)\mathbf{i} \qquad (6)$$

$$F_y\mathbf{j} + mg\cos\theta\sin\phi\mathbf{j} = \delta m\left(\frac{dv_y}{dt} + v_x\omega_z - v_z\omega_x\right)\mathbf{j} \qquad (7)$$

$$N\mathbf{k} - \cos\theta\cos\phi mg\mathbf{k} = \delta m\left(\frac{dv_z}{dt} + v_y\omega_x - v_x\omega_y\right)\mathbf{k} \qquad (8)$$

$$\begin{aligned}I_x\alpha_x - (I_y - I_z)\omega_y\omega_z &= 0\\ I_y\alpha_y - (I_z - I_x)\omega_x\omega_z &= 0\end{aligned} \qquad (9)$$

$$\begin{aligned}I_z\alpha_z\mathbf{k} - (I_x - I_y)\omega_x\omega_y\mathbf{k} &= M_z\mathbf{k}\\ &= (F_{t2} - F_{t1})\mathbf{i} \times \frac{B}{2}\mathbf{j} - M_r\mathbf{k}\\ &= (F_{t2} - F_{t1})\frac{B}{2}\mathbf{k} - M_r\mathbf{k}\end{aligned} \qquad (10)$$

in (6), (7) and (8):

F_f—deformed resistance of ground;

m—mass of mobile robot;

δ—Gain coefficient of mass, which converts the rotational inertia equally to translational mass, dimensionless. For a general mechanical system with some rotational components, in general, $\delta = 1.1 \sim 1.2$ [9].

F_y: forces in y axis direction, in fact, it is a centripetal force;

And in (9) and (10):

ω_z: angular velocity of UGV relative to axis z;

α_z: angular acceleration of UGV relative to axis z;

I_z : rotational inertia of UGV according to axis z;

M_r: resistant torque applied on UGV;

B: distance between the two centres of tracks.

The deformed resistance of ground, F_f is the function of normal pressure of ground

$$F_f = fN_z \qquad (11)$$

Where f denotes the coefficient of deformed resistance. N_z is the dot product of vector of gravity and the normal vector of contact area

$$N_z = -mg\mathbf{K} \cdot \mathbf{k} \qquad (12)$$

According to [10], state equations of the attitude angles of UGV are

$$\begin{bmatrix} \omega_x \\ \omega_y \\ \omega_z \end{bmatrix} = \begin{bmatrix} -\sin\theta & 0 & 1 \\ \sin\phi\cos\theta & \cos\phi & 0 \\ \cos\phi\cos\theta & -\sin\phi & 0 \end{bmatrix} \begin{bmatrix} \dot{\psi} \\ \dot{\theta} \\ \dot{\phi} \end{bmatrix} \quad (13)$$

Resistant torque, M_r, is the function of mg, L, B and the attitude angles of UGV, which will be discussed in the following section.

3 Disturbances on UGV

When a UGV is running outdoors, many kinds of disturbances coming from slope and rugged surface will affect its motion status, and furthermore, on exceeding some limits, motion stability of UGV will be challenged immensely. Further research reveals that there are three types of disturbances according to the mechanism of disturbances: (1) centrifugal force, (2) slope force and (3) general disturbances on yaw motion.

As the affection of centrifugal force yielded minor effect on the UGV, therefore, only effects of slope force and general yaw motion were discussed here.

Slope force is a reacting force, which is to counteract the effect of the gravity driving UGV down slope. As in this research, the UGV mainly works in 3-D space; therefore, the slope force cannot be ignored.

General disturbance on yaw motion of UGV is the abstract disturbance of several kinds of disturbances. If disturbances are applied on the two tracks of the UGV equally to, then, the yaw angle will not be affected and the UGV can keep stable in the direction of motion. Unfortunately, UGV seems never so lucky to get balanced loads on two tracks. Difference of the coherent status of the two tracks, difference of the shape of terrain under the two tracks, as well as random lateral resistant force, can yield additional disturbances on the yaw motion of the UGV, which have the same feature that all can be denoted as the disturbances on the angular acceleration.

Before going into detailed discussion, the following hypotheses are necessary:

The surface of ground is large enough to assure the contacting surface of tracks are local planar, hence, in spite of the variation of curvature of ground, the local area where UGV locates is considered as local plane; Suppose that lateral motion of UGV is in steady state, without any acceleration.

Effects of slope force are not only in the lateral direction of UGV but also in the longitudinal direction, and the magnitude is time-variable:

$$F_t = m\cos\theta\sin\phi \quad (14)$$

Accordingly, incited offset can be yielded:

$$OF = \frac{m\cos\theta\sin\phi L}{2\mu_t N_z} \quad (15)$$

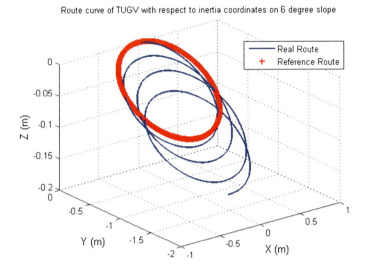

Fig. 2 Space curve of UGV on slope $\phi = 6°$

and steady lateral velocity

$$v_y = \omega_z \times OF$$
$$= \frac{m\omega_z \cos\theta \sin\phi L}{2\mu_t N_z} \qquad (16)$$

As the F_t also incites the rearrangement of lateral resistant shearing stress, then:

$$M_r = \frac{\mu_t mgL \cos\theta \cos\phi}{4} + \frac{m\omega_z^2 \cos\theta \sin\phi \tan\phi L}{4\mu_t g} \qquad (17)$$

From (16), if UGV doesn't rotate, then, no v_y appears, which means that the slope force doesn't affect the stability of straight-line motion. Consequently, the behaviours of UGV in two turning processes under the disturbance of slope force are investigated in MATLAB/Simulink, one is started from the initial condition of $\phi = \pi/60(3°), \theta = \psi = 0$, and the other is started from the initial condition of $\phi = \pi/30(6°), \theta = \psi = 0$. Figure 2 illustrates the results of $\phi = 6°$ ($n_{m1} = 3000r/\text{Min}, n_{m2} = 1000/r\text{Min}$) and Fig. 3 illustrates the results of $\phi = 3°$ ($n_{m1} = 3000r/\text{Min}, n_{m2} = 1000/r\text{Min}$).

Disturbances mentioned above are connatural for UGV in 3-D space, while some random disturbances in motion also make UGV deviate from stable state. For example, when a UGV is running on rough terrain, if track in one side is running over a flat surface while track on the other side has to overcome a small obstacle with the same speed, then in fact, the UGV undertakes a small turning in this procedure in yaw direction; another probable situation is that UGV is running on a planar surface but there are two diverse contacting conditions under the two tracks, for instance, one is slippery and the other is solid, in which situation the

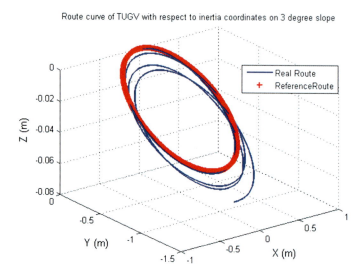

Fig. 3 Space curve of UGV on slope $\phi = 3°$

UGV leads to turn to the direction of slippery side, then, unexpected yaw motion occurs. Definitely, there are many other disturbances which will make UGV rotate in yaw direction with vastly different disturbing mechanism. It is a great challenge to establish mechanics model for every disturbance. On analysing the behaviour of these disturbances, the most common feature is the result of disturbances can be reflected by the rate of variation of angular velocity in axis z (i.e. angular acceleration in axis z). Therefore, in this research, all of disturbances in this conditions are converted into the noise angular acceleration in axis z.

To clarify the effects of general yaw motion disturbance, simulation was carried out, in which UGV was supposed to follow a straight line coincided with axis x in inertia coordinates. The disturbance applied on UGV is illustrated in Fig. 4, and the effects on route of UGV are shown in Fig. 5.

Discussions:

From figures above, the effects of slope force are significant: the real route of UGV is obviously deviated from the reference route; and when UGV is subjected to the general yaw motion disturbance, real route will deviate from predefined route rapidly.

4 Design of PID Based Motion Stabilizing Controller of UGV

Based on the discussion above, to keep the motion stability of UGV, all of disturbances should be compensated by a controller appropriately, which is named as stabilizing controller. Starting from convenience and simplicity, the first choice for

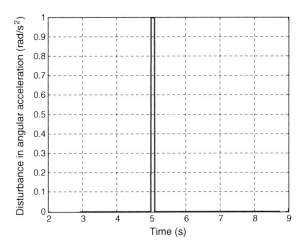

Fig. 4 Disturbance in angular acceleration

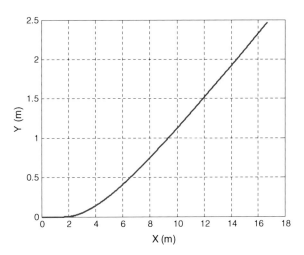

Fig. 5 Disturbed route of UGV

attempt is employing a PID controller, which is the most universal but powerful tool for control engineering.

There are two necessary functions or modules which should be realized by the designed stabilizing controller:

Motion control module (path following and/or attitude control): Firstly, stabilizing controller accepts the instructions from planning or navigation module of UGV, which are the descriptions on status (path and/or attitude) UGV should be in; and in general; the path curve is differentiable. Secondly, on receiving data of motion sensors, stabilizing controller manages to map the path and/or attitude information into expected rotational speed of motors for driving; and finally, transfer the motion instructions to the controller of motors to complete a control cycle.

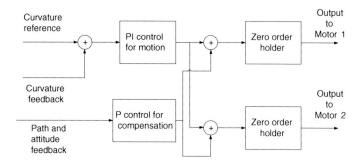

Fig. 6 Schema of PID controller

Compensation module based on estimation of disturbances: Obviously, effects of disturbances can be represented in the variation of parameters of motion (such as curvature of path), and can be adapted by the motion control module. But, as the control system and driving system must take a short period of time to response the disturbances passively, therefore, stability performance will be depraved. Thus, active compensations based on the estimation of disturbances are preferred to improve the efficiency of stabilizing controller.

Integration of motion control module and compensation module: Based on the modules mentioned above, an integration module should be presented to compose the final control output advisably.

As the most extensive control algorithm, PID may be the first choice for all kinds of control issues, as well as in mobile robot [11–15]. In this research, a digital PI controller is employed as a motion control module, and a digital P controller is for the compensation module, which is illustrated in Fig. 6.

In this project, there is no direct information of position for motion controller, and the shape of path will be the key parameter. In the motion control module, the curvature of the path is selected to be the path reference input, which is most convenient for representing some specific curves utilized in path planning. For instance, the curvature of a straight line is 0 and that of a circle with radius of r is $1/r$, and other complex curves can be presented by the combination of straight line and parts of circle.

Correspondingly, the discrete PID controller is a direct conversion from continuous PID controller, which can be expressed as follows

$$G(z) = K_p \left(1 + \frac{Tz}{T_i(z-1)} + \frac{T_d(z-1)}{Tz} \right) \qquad (18)$$

where T is the sampling time.

In compensation module, a P controller is designed as

$$Output = K_\phi \omega_z \phi \cos 2\phi \qquad (19)$$

Motion Stabilizing Controller of Off-Road Unmanned Wheel Vehicle

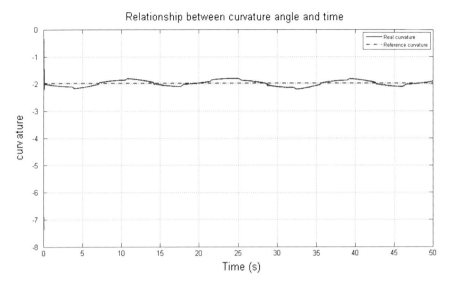

Fig. 7 Comparison of real curvature and reference curvature on 3° slope (*left above*)

where

K_ϕ = proportional gain of inclined slope
ϕ = inclined angle of slope
The implemented controller is shown in Fig. 6.

5 Simulation Results Under the Control of PID Controller

On designed the two modules, simulations are carried out. Firstly, situation under disturbance incited by slope force is executed. Figures 7 and 8 show the control results on slope of 3°, and Figs. 9 and 10 present the control results on slope of 6°.

Secondly, simulations under the general disturbance of yaw motion on plane surface are put forward with the same disturbing noise illustrated in Fig. 4 and the results are shown from Figs. 11 and 12.

Discussions:

(1) For a PID controller (PI plus P control in fact), the control effects on motion stabilizing are significant (compared with Figs. 2 and 10, or Figs. 3 and 8). When slope angle is 3°, the maximum deviation between expected curve and real curve is 0.0211 m under the PI control, while the that of situation without control is 0.2510 m; when slope angle is 6°, the maximum deviation between expected curve and real curve is 0.1215 m under the PI control, while the that of situation without control is 0.5836 m.

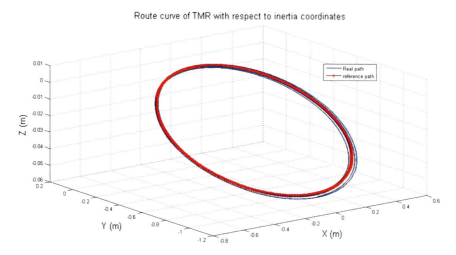

Fig. 8 Comparison of real path and reference path on 3° slope (*right above*)

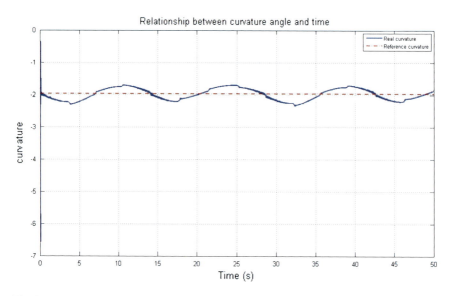

Fig. 9 Comparison of real curvature and reference curvature on 6° slope (*left above*)

(2) The control effects under the general disturbance of yaw motion are interesting. On one hand, the controller did correct or compensate the disturbance on the angular velocity (the deviation after 50 s is −0.44 m; while that without control is 2.44 m and become infinite as time is elapsing); on the other hand, it overdoes in some degree and the new steady state of UGV deviates from the origin of yaw motion obviously (Fig. 11).

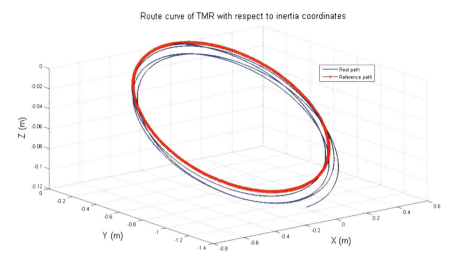

Fig. 10 Comparison of real path and reference path on 6° slope (*right above*)

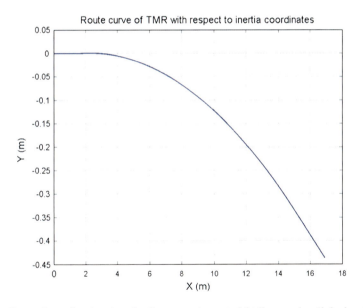

Fig. 11 Comparison of real path and reference path on straight-line running (*left above*)

Above all, the motion stabilizing controller based on PID approach can achieve fairish effects, but it seems that some intelligence is necessary to compensate the general disturbance in yaw motion.

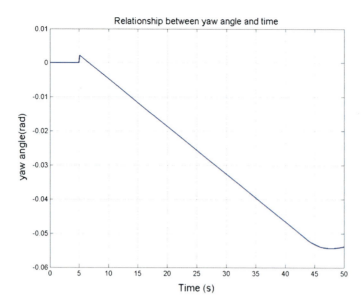

Fig. 12 Relationship of yaw angle ψ and time (*right above*)

6 Conclusions

In this paper, the mechanisms of two kinds of disturbances applied on UGV have been thoroughly discussed, which reveal the deviational rules of UGV. With the simulated case studies, disturbances of slope force and general disturbing on yaw motion represent characteristics individually.

Preceded from practical view of point, a PID controller (PD plus P control, in fact) is realized to keep the stability of motion for UGV. Simulation results prove the validity for disturbance incited by slope force, but also reveal less effectiveness for general disturbance on yaw motion, which will be resolved in the future research.

References

1. Bekey GA (2005) Autonomous robots: from biological inspiration to implementation and control. MIT Press, Cambridge
2. Lyapunov AM (1966) Stability of motion. Academic Press, London
3. Faldin NV, Morzhov AV, Boiko IM (2007) Stability of periodic motions in relay feedback systems with saturation in plant dynamics. In: American control conference, 2007. ACC '07, pp 4721–4726
4. Shibata T, Murakami T (2008) Null space motion control by PID control considering passivity in redundant manipulator. IEEE Trans Ind Inform 4:261–270

5. Zhao J, Yao Y (2006) On the sudden change in joint velocity during fault tolerant operations for spatial coordinating redundant manipulators. IEEE/RSJ International conference on intelligent robots and systems, 2006, pp 3885–3890
6. Kin K, Yano O, Urabe H (2003) Enhancements in vehicle stability and steerability with slip control. JSAE Rev 24:71–79
7. Shino M, Nagai M (2003) Independent wheel torque control of small-scale electric vehicle for handling and stability improvement. JSAE Rev 24:449–456
8. Miller TJE (1989) Brushless permanent-magnet and reluctance motor drives. Clarendon Press, Oxford
9. Ma Y, Zhu QM, Winfield A, Yan QD (2009) Simulation research on braking performance of hydrodynamic torque converter and retarder based on automatic shifting rules. Int J Modell Identif Control 8:80–87
10. Baruh H (1999) Analytical dynamics. WCB McGraw-Hill, Boston
11. Gong Z, Guzman JI, S. S. J., R. David C, Dissanayake G, and Hugh (2004) A heuristic rule-based switching and adaptive PID controller for a large autonomous tracked vehicle: from development to implementation. In: Proceedings of the 2004 IEEE international conference on control applications Taipei, Taiwan, vol 8633, pp 1272–1277
12. Normey-Rico JE, Alcalá I, Gómez-Ortega J, Camacho EF (2001) Mobile robot path tracking using a robust PID controller. Control Eng Practice 9:1209–1214
13. Hogg RW, Rankin AL, Roumeliotis SI (2002) Algorithms and sensors for small robot path following. In: Proceedings ICRA 02, IEEE international conference on robotics and automation, 2002, pp 3850–3857
14. Rena T-J, Chen T-C, Chen C-J (2008) Motion control for a two-wheeled vehicle using a self-tuning PID controller. Control Eng Practice 16:365–375
15. Ye J (2008) Adaptive control of nonlinear PID-based analog neural networks for a nonholonomic mobile robot. Neurocomputing 71:1561–1565

Printed by Publishers' Graphics LLC
BT20121218.10.02.129